Systems Biology Application in Synthetic Biology

Shailza Singh
Editor

Systems Biology Application in Synthetic Biology

Springer

Editor
Shailza Singh
Computational and Systems Biology Lab
National Centre for Cell Science
Pune, India

ISBN 978-81-322-2807-3 ISBN 978-81-322-2809-7 (eBook)
DOI 10.1007/978-81-322-2809-7

Library of Congress Control Number: 2016952540

Printed on acid-free paper

This Springer imprint is published by Springer Nature
The registered company is Springer (India) Pvt. Ltd.

Preface

Systems and synthetic biology is an investigative and constructive means of understanding the complexities of biology. Discovery of restriction nucleases by Werner Arber, Hamilton Smith, and Daniel Nathans in 1978 revolutionized the way DNA recombinant constructs were made and how individual genes were analyzed for its function and vitality. It also opened the doors to a new era of "synthetic biology" where apart from analysis and description of existing gene, new gene arrangements can be constructed and evaluated. Since then, synthetic biology has emerged from biology as a distinct discipline that quantifies the dynamic physiological processes in the cell in response to a stimulus. Switches, oscillators, digital logic gates, filters, modular – interoperable memory devices, counters, sensors, and protein scaffolds are some of the classic design principles based on which many more novel synthetic gene circuits can be created with possible application in biosensors, biofuels, disease diagnostics, and therapies. Most of these gene networks combine one or more classes of controller components, such as conditional DNA-binding proteins, induced-protein dimerization, RNA controllers, and rewired cell-surface receptors, to modulate transcription and translation that alters protein function and stability.

An iterative design cycle involving molecular and computational biology tools can be capitalized to assemble designer devices from standardized biological components with predictable functions. Research efforts are priming a variety of synthetic biology inspired biomedical applications that have the potential to revolutionize drug discovery and delivery technologies as well as treatment strategies for infectious diseases and metabolic disorders. The building of complex systems from the interconnection of parts or devices can be significantly facilitated by using a forward-engineering where various designs are first optimized, tested *in silico* and their properties are assessed using mathematical analysis and model-based computer simulations. Mathematical models using Ordinary Differential Equations (ODEs), Partial Differential Equations (PDEs), Stochastic Differential Equations (SDEs), or Markov Jump Processes (MJPs) are typically used to model simple synthetic biology circuits. Thus use of computation in synthetic biology can lead us to ways that help integrate systems models to support experimental design and engineering. Synthetic biology has significantly advanced our understanding of complex control dynamics that program living systems. The field is now starting to tackle relevant therapeutic challenges and provide novel diagnostic tools as well as unmatched therapeutic strategies for treating significant human pathologies. Although synthetic biology-inspired treatment concepts are still far from being applied to any licensed drug or therapy, they are rapidly developing toward clinical trials. Nevertheless, it has provided insights into disorders that are related to deficiencies of the immune system known for its complex control circuits and interaction networks.

Novel-biological mechanism may also be coupled with image-modeling approach to be verified in *in vitro* conditions. Computational techniques can be used in tandem with image analysis to optimally characterize mammalian cells, leading to results that may allow scientists to uncover mechanisms on a wide range of spatio-temporal scales. These elucidated methods and principles used in *in silico* hypotheses generation and testing have the potential to catalyze discovery at the bench. Despite considerable progress in computational cell phenotyping, significant obstacles remain with the magnitude of complexity with experimental validation at the bench. The true power of computational cell phenotyping lies in their strengths to generate insights toward *in vivo* constructs, which is a prerequisite for continued advancements. None of the obstacles is insurmountable. However, advances in imaging and image processing may transcend current limitations which may unlock a wellspring of biological understanding, paving the way to novel hypotheses, targeted therapies, and new drugs. Additionally, phenotyping permits the effects of compounds on cells to be visualized immediately without prior knowledge of target specificity. By harnessing the wealth of quantitative information embedded in images of *in vitro* cellular assays, HCA/HCS provides an automated and unbiased method for high-throughput investigation of physiologically relevant cellular responses that is clearly an improvement over HTS methods, allowing significant time and cost savings for biopharmaceutical companies. The emergence of non-reductionist systems biology aids in drug discovery program with an aim to restore the pathological networks. Unbalance reductionism of the analytical approaches and drug resistance are some of the core conceptual flaws hampering drug discovery. Another area developing and envisaged in this book is system toxicology, which involves the input of data into computer modeling techniques and use differential equations, network models, or cellular automata theory. The input data may be biological information from organisms exposed to pollutants. These inputs are data mostly from the "omics," or traditional biochemical or physiological effects data. The input data must also include environmental chemistry data sets and quantitative information on ecosystems so that geochemistry, toxicology, and ecology are modeled together. The outputs could include complex descriptions of how organisms and ecosystems respond to chemicals or other pollutants and their inter-relationships with the many other environmental variables involved.

The model outputs could be at the cellular, organ, organism, or ecosystem level. Systems toxicology is potentially a very powerful tool, but a number of practical issues remain to be resolved such as the creation and quality assurance of databases for environmental pollutants and their effects, as well as user-friendly software that uses ecological or ecotoxicological parameters and terminology. Cheminformatics and computational tools are discussed in lengths which help identify potential risks including approaches for building quantitative structure activity relationships using information about molecular descriptors. The assimilation of chapter from various disciplines includes the trade-offs and considerations involved in selecting and using plant and other genetically engineered crops. Systems biology also aid in understanding of plant metabolism, expression, and regulatory networks. Synthetic biology approaches could benefit utilizing plant and bacterial "omics" as a source for the design and development of biological modules for the improvement of plant stress tolerance and crop production. Key engineering principles, genetic parts, and computational tools that can be utilized in plant synthetic biology are emphasized.

The collection of chapters represents the first systematic efforts to demonstrate all the different facets of systems biology application in synthetic biology field.

I would like to thank Mamta Kapila, Raman Shukla, Magesh Karthick Sundaramoorthy, and Springer Publishing group for their assistance and commitment in getting this book ready for publication. I would also like to thank my wonderful graduate students Vineetha, Milsee, Pruthvi, Ritika, Bhavnita, and Dipali for being a rigorous support in the entire endeavor. Finally, I would especially like to thank my family, Isha and Akshaya, my parents for being patient with me during the process. Without their love and support, this book would not have been possible.

Pune, India Shailza Singh

Contents

About the Editor

Shailza Singh is working as Scientist D at National Centre for Cell Science, Pune. She works in the field of Computational and Systems Biology wherein she is trying to integrate the action of regulatory circuits, cross-talk between pathways and the non-linear kinetics of biochemical processes through mathematical models. The current thrust in her laboratory is to explore the possibility of network-based drug design and how rationalized therapies may benefit from Systems Biology. She is the recipient of RGYI (DBT), DST-Young Scientist and INSA (Bilateral Exchange Programme). She is the reviewer of various international and national grants funded from government organizations.

1 Microbial Chassis Assisting Retrosynthesis

Milsee Mol, Vineetha Mandlik, and Shailza Singh

1.1 Introduction

It's a well-known and a documented fact that life has arisen from simple molecules. Therefore the main stay of research in biology is to strip down the inherent complexity associated due to the interaction between these simple molecular assemblies. During the course of evolution, there has been a reduction in the complexity that constitutes the essential features of a living cell. The comprehension (if it is possible to comprehend fully) of the underlying complexities will not only allow us to understand the key regulatory mechanism in numerous diseases, production of important metabolites, etc. but also help us to build a reliable mathematical model for formulating future scientific enquiry. A better understanding of cellular systems can be done via two competing routes the "bottom up" as well as "top-down" synthetic biology approach. Synthetic biology has two goals: to re-engineer existing systems for better quantitative understanding; and, based on this understanding engineer new systems that do not exist in nature [1]. The fundamental principle of synthetic biology is similar to constructing non-biological system e.g. a computer, by putting together composite, well-characterized modular parts. It is an interdisciplinary science drawing expertise from biology, chemistry, physics, computer science, mathematics and engineering [2].

Synthetic biology has re-revolutionized the way biology is done today in laboratories across the globe, also mainly because of the way DNA the blue print of a cells functionality is being synthetized by simply providing the desired sequence to the automated synthesizer. Synthetic biologists are now on the verge of developing 'artificial life' that has enormous applications in biotechnology apart from the fact that it is being used to now understand the origin of life. The 'top-down' approaches in synthetic biology are being used to synthesize the minimal cells by systematically reducing the genome of a cell such that it shows a desired function under environmentally favourable conditions [3, 4]. Successful chemical synthesis of genome and its transfer to the bacterial cytoplasm [5] reveals the power of synthetic biology framework to create a minimal cell for greater application in biotechnology [2]. Such a minimal cell having the minimum required genome could serve as a "chassis" that can be further expanded with the addition of genes for specific functions desired from a tailor-made organism. Further a streamlined chassis based on a minimal genome can simplify the interaction

M. Mol • V. Mandlik
National Centre for Cell Science, NCCS Complex, Ganeshkhind, Pune 411007, India

S. Singh (✉)
Computational and Systems Biology Lab, National Centre for Cell Science, NCCS Complex, Ganeshkhind, Pune 411007, India
e-mail: shailza_iitd@yahoo.com; singhs@nccs.res.in

S. Singh (ed.), *Systems Biology Application in Synthetic Biology*,
DOI 10.1007/978-81-322-2809-7_1

between the host and the system that may have relevance in minimizing the effect of the metabolic burden of the exogenous pathway placed in the cell [6]. Such extensively streamlining is possible for many of the medically and industrially important microorganism as their genomes have been already sequenced and assembled.

Comparative genomics is a useful methodology that delineates genes based on the conservedness of the genes to distant related species. It is based on the hypothesis that the conserved genes are certainly essential for cellular function and may be well approximated to the required minimal gene set [7]. But as more and more genomes are being sequenced there is divergence in the evolutionary tree showing that some of the essential functions can be performed by non-orthologous genes [8]. Therefore, gene persistence rather than gene essentiality should be taken into consideration for constructive way to identify the minimal universal functions supporting robust cellular life [9].

Another approach that of experimental gene inactivation identifies genes, those are important for the viability of the cell. Genome-scale identifications of such genes have been done using the prokaryotic as well as eukaryotic systems using strategies of massive transposon mutagenesis [10, 11], the use of antisense RNA [12] to inhibit gene expression and the systematic inactivation of each individual gene present in a genome [13, 14]. These genome scale identifications have been done under predefined experimental growth conditions. This kind of experimental identification helps us get a complete understanding of the relationship between genotype and phenotype which would facilitate the design of minimal cell [8].

The data generated in such genome scale experimental models is large which needs computer-assisted mathematical treatment to get some meaningful statistically valid approximations. Therefore mathematical models that relate the gene content (genotype) of a cell to its physiological state (phenotype) enables the simulation of minimal gene sets under various environmental growth conditions (constraint-based approach) [15–18]. Thus, *in silico*, with in the complex gene network reaction(s), each gene can be individually "deleted" (flux 'zero') and relate it to the biomass as the fitness function for the system [19]. This flux-based models yield key evolutionary insights on the minimal genome [20].

Integrating all the information from comparative genomics, experimentation and *in silico* predictions, a new approach of retrosynthesis is rising for building *de novo* pathways in host chassis [21–23]. Retrosynthesis is a technique routinely used in synthetic organic chemistry [24, 25], where it starts by conceptually defining the structure and properties of the desired molecule to be produced and working backward through known chemical transformations to identify a suitable precursor or sets of precursors. This approach when applied to biological metabolic transformation can identify the reactions involved and their corresponding enzymes. Thus, by enumerating the biochemical pathways, it can be linked to the final product in the host's metabolism [23].

With the available tool kits for designing biological systems, the future predictability is relatively difficult and may lead to bottleneck situation in the production pipeline. Metabolic pathway models are being made more predictable by incorporating the freedom to tweak the gene expression to achieve a particular flux of each metabolite in the reaction or pathway [26]. Tools that help in debugging bottleneck in the metabolic pathway would reduce development times for optimizing engineered cells. Functional genomic tools can serve this purpose [27], which helps in chalking out the over or under production of a protein/enzyme in the pathway that can lead to a stress response [28, 29]. The information from these tools can be rendered to diagnose the problem and modify expression of genes in the metabolic pathway to improve productivity. Taking advantage of the cell's native stress response pathways, too many desirable chemicals particularly at the high titres needed for industrial-scale production can be an effective way to overcome product toxicity [30].

1.2 Tools for Designing and Optimizing Synthetic Pathway

It is an uphill task to find an optimal solution for a selected pathway, enzymes or chassis organism from an abundance of possibilities. Engineering a synthetic pathway and uploading it into the chassis organism followed by optimizing the production of the desired product involves lot of experimental work which is accompanied by lots of permutation and combinations of conditions. To make life easy for a synthetic biologist powerful computational tools are a necessity. There are many computational tools that can lead for a better informed, rapid design and implementation of novel pathways in a selected host organism with the desired parts and flux of the desired product is listed in Table 1.1. These tools are based on criteria like pathway selection and thereafter ranking them. These prediction help to explore the pathways that are chemically versatile and also help compare their efficiencies as compared to the natural pathways. Organism selection for uploading the novel pathway depends on two approaches: First, choose an organism that already has most of the reactions involved in the pathway, thereby reducing the stochasticity that can be introduced due to the new enzymes in the metabolic network [23]. The second approach is to build genome scale models using constraint-based flux balance analysis. In this approach, steady-state flux distribution of the metabolic network is predicted based on the stoichiometry of each reaction, mass–balance constraints and an objective function specifying the fluxes of components that are to be optimized [31]. Once the prioritized pathway and optimum host is selected, the next step is to construct the pathway by using parts such as the RBS, promoters, terminators, etc. with the regulatory elements incorporated. A range of standardized and characterized parts are available at the parts registry [32]. Efforts are underway to increase the catalogue available at the registry, as they are suitable for finding regulatory elements rather than the coding sequences. Since the coding sequences for the enzymes are part of a specific synthetic pathway, they are not catalogued and for this purpose genome-mining is a crucial step. The last part of the process design is to synthesize the DNA parts that are codon optimized for the host chassis. Many variants of the basic DNA sequence can also be synthesized from which an efficient sequence can be picked up. After all the above steps are succesfully completed a functional design can be arrived to, which can then be inserted into the chromosome of the host genome [33] or as a multigene expression plasmid [34]. The workflow designing a synthetic pathway into a microbial chassis system can be depicted pictorial in Fig. 1.1.

1.3 Choosing a Host and Vector for Synthetic Pathway Construction

Choosing a correct heterologous host for the production of a desired product is an important and uphill task in metabolic engineering of microbes. A host must be chosen based on the fact whether the desired metabolic pathway already exists or can it be reconstituted in that host. If so, then the host can survive under the desired process conditions of pH, temperature, ionic strength, etc. for the optimum titre of the desired product. The host should be genetically robust and should not be susceptible to phage attacks and at the same time should be amenable to available genetic tools. Although *E. coli* can be treated with different genetic tools available, it has disadvantage of being susceptible to phage attack. The host should be able to grow on simple, inexpensive carbon sources without or with minimal additions to the process media, thereby reducing the production cost of the product [63, 64]. Another aspect that should be considered is the level of expression of the heterologous enzymes in the host strain. The enzymes should be expressed in amounts that are catalytically important for the conversion of the starting material to the desired product. Toxicity of the intermediate metabolites for the hosts should also be dealt with, because any intermediate that is toxic will have a profound effect on the final titres of the desired product.

Table 1.1 Computational tools currently being employed for synthetic pathway construction

Tool		Description
Pathway prediction	BNICE (Biochemical Network Integrated Computational Explorer) [35]	Identification of possible pathways for the degradation or production of a desired compound within a thermodynamic purview
	DESHARKY [36]	Best match pathway identification specific to a host; provides phylogenetically related enzymes
	RetroPath [37]	Retrosynthetic pathway design, pathway prioritization, host compatibility prediction, toxicity prediction and metabolic modelling
	FMM (From Metabolite to Metabolite) [38]	Finds an alternate biosynthetic routes between two metabolites within the KEGG database
	OptStrain [39]	Optimization of the host's metabolic network by suggesting addition or deletion of a reaction
Parts identification	Standard Biological Parts knowledgebase [40]	Knowledgebase with parts for easy computation; includes all the parts from Registry of Standard Biological Parts
	IMG (Integrated Microbial Genomes) [41]	Comparative and evolutionary analysis of microbial genomes, gene neighbourhood orthology searches
	antiSMASH [42]	Identification, annotation and comparative analysis of secondary metabolite biosynthesis gene clusters
	KEGG [43]	Database of organism specific collection of metabolite and metabolic pathway
Parts optimization and synthesis	RBS Calculator [44]	Automated design of RBSs based on a thermodynamic model of transcription initiation
	RBSDesigner [45]	Algorithm for prediction of mRNA translation efficiencies
	Gene Designer 2.0 [46], Optimizer [47],	Gene, operon and vector design, codon optimization and primer design
	DNAWorks [48], TmPrime [49]	Oligonucleotide design for PCR-based gene synthesis, with integrated codon optimization
	CloneQC [50]	Quality of sequenced clones by detecting errors in DNA synthesis
Pathway and circuit design	Biojade [51]	Software tool for design and simulation of genetic circuits
	Clotho [52]	Flexible interface for synthetic biological systems design; within the interface, a range of apps/plugins can be utilized to import, view, edit and share DNA parts and system designs
	GenoCAD [53]	CAD software that allows drag-and-drop drawing and simulation of biological systems
	Asmparts [54]	Computational tool that generates models of biological systems by assembling models of parts
	SynBioSS [55]	Designing, modelling and simulating synthetic genetic constructs
	CellDesigner [56]	Graphical drawing of regulatory and biochemical networks that can be stored in Systems Biology Markup Language (SBML)
Metabolic modelling	COBRA Toolbox [57]	Metabolic modelling and FBA
	SurreyFBA [58]	Constraint-based modelling of genome-scale networks
	CycSim [59], BioMet Toolbox [60]	Analysing genome-scale metabolic models; includes enzyme knockout simulations
	iPATH2 [61], GLAMM (genome-linked application for metabolic maps) [62]	Interactive visualization of data on metabolic pathways

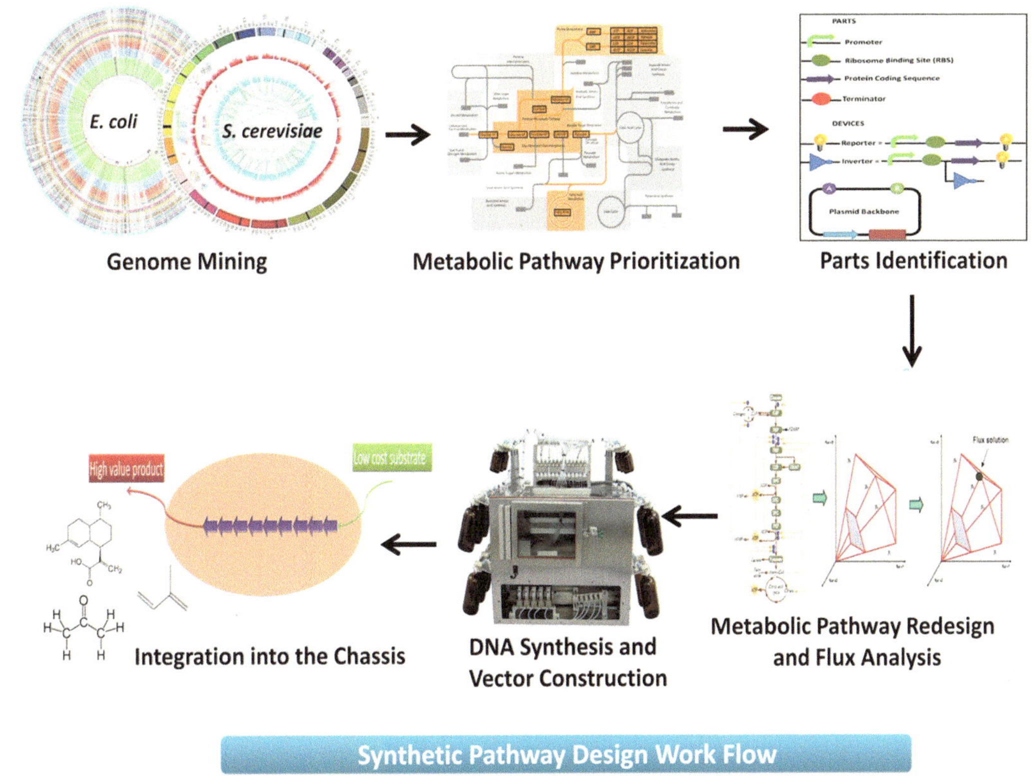

Fig. 1.1 Synthetic pathway design workflow

All the genetic manipulations involve the construction of a vector that contains all the enzymes required to reconstitute the novel metabolic pathway in the heterologous host. Therefore the cloning vector should be stable, have a consistent copy number, should replicate and express large sequences of DNA. The enzyme production rate from these vectors can be tuned to the desired levels by varying the promoter [65], ribosome binding strength [66] and stabilizing the half-life of the mRNA [67]. Of these, promoters are essential in controlling biosynthetic pathways that respond to a change in growth condition or to an important intermediary metabolite [68, 69]. These kinds of promoters allow inexpensive and inducer-free gene expression. Once a vector with all the desired properties is constructed the expression of the genes should be well coordinated, which can be done using a non-native RNA polymerase or transcription factor that can induce multiple promoters [70]; group related genes into operons; vary ribosome binding strength for the enzymes encoded in the operon [71]; or controlling mRNA stability of each coding region [72].

1.4 Important Breakthrough in Metabolic Engineering Using Synthetic Biology Approach

Though synthetic biology and construction of unnatural pathways is in its infancy, several pioneering experimental efforts in this direction have highlighted the immense potential of the field. In parallel, DNA sequencing has revealed a huge amount of information within the cellular level in terms of isozymes catalysing the same reaction in different organism. Alongside development of curated databases for the reaction catalysed by these enzymes are aiding the discovery of novel routes for pathway reconstruction in heterologous host chassis organisms such as *E. coli*, *Saccharomyces cerevisiae*, *Bacillus subtilis* and *Streptomyces coelicolor*. These organisms are amenable to the new genetic tools that enable more precise control of the reconstructed metabolic pathways. Newer analytical tools that enable track RNA, protein and metabolic intermediates can help identify rate limiting kinetic reactions in the pathway that helps design novel recombinant enzymes [68].

Many natural pathways can be transferred to the microbial chassis for the production of natural chemicals originally synthesised by plants and whose chemical synthesis is complex or expensive. These pathways are important as they are source to important natural molecules like alkaloids, polyketides, nonribosomal peptides (NRPs) and isoprenoids that find their application in pharmaceuticals. Similarly, fine chemicals such as amino acids, organic acids, vitamins and flavours have been produced economically from engineered microorganisms [68].

One of the most notable examples is that of artemisinin, a potent antimalaria drug produced naturally in plant *Artemisia annua*. Large-scale production of this compound is costly and varies seasonally. To overcome these practical challenges, synthetic biologists have engineered its yeast-derived biosynthetic pathway (isoprenoid precursor) in the bacterium *Escherichia coli* [73]. Later, a synthetic pathway consisting dual enzyme origin (plant- and microorganism) capable of producing artemisinic acid that can be converted into artemisinin in just two chemical steps was installed in *E. coli* and *Saccharomyces cerevisiae* [74–76]. The titre of artemisinic acid was high compared to the titres achieved from its natural plant source. Another plant-derived pathway to produce taxadine, which is the first committed intermediate for the anticancer drug taxol, was successfully introduced in *E. coli*. After careful balancing of the expression of the heterologous pathway and the native pathway producing the necessary isoprenoid precursors, more than 10,000-fold production level was achieved [77]. An important building block d-hydroxyphenylglycine for the side chain of semi-synthetic penicillins and cephalosporins was also synthesized using the workflow of synthetic pathway design. It was done by combining enzymes hydroxymandelate synthase from *Streptomyces coelicolor*, hydroxymandelate oxidase from *Amycolatopsis orientalis* and hydroxyphenylglycine aminotransferase from *Pseudomonas putida* [78]. Synthetic circuits are also designed in integration with the host metabolic pathway for the controlled release of therapeutic *in situ*. Devices that sense pathogenic conditions such as cancer cells, pathogenic microorganisms and metabolic states are designed to fine-tune transgene expression in response to these conditions [79–81]. These sensors could be small molecules as autoinducers to light sensitive devices [82] and miRNA detection systems [83]. A refined circuit was developed for that could sense hyperuricemic condition associated with the tumour lysis syndrome and gout [84].

Biofuel namely isopropanol and higher alcohols was re-routed in the native metabolism in *E. coli*, by combining enzymes from various biological sources [81, 82] Elaborate synthetic approaches have redesigned specific transcriptional regulatory circuits with combination of enzymes from other microorganisms that led to the production of biodiesels and waxes from simple sugars in *E. coli* [83]. In the synthesis of methyl halides from 89 putative homologues of the enzyme methyl halide transferase from bacteria, plants, fungi and archaea were identified by a BLAST search. All the retrieved homologues were codon-optimized to be expressed in *E. coli*. The codon-optimization led to build a synthetic gene library, which was tested for optimum desired function in the host strain, resulting in high production titres of methyl halide [84]. Similarly microbial biofuel export and tolerance was enhanced by creating a synthetic library of hydrophobe/amphiphile efflux transporters [85].

As the engineering aims become more ambitious, a trend towards more prominent application of synthetic pathway design and implementation will lead to increased efficiency and may also incorporate more complex metabolic pathways.

1.5 Future Applications

Bulk chemicals such as solvents and polymer precursors are all produced through chemical catalysis from petroleum. The dwindling reserves and trade imbalances in the petroleum market and low-cost production of these bulk chemicals can be an avenue for the application of microbial engineering from starting material like starch, sucrose or cellulosic biomass [68]. The process pipeline for production of petroleum based transportation fuel is expensive but at the same time it is the most valued product in the world. Engineered biological systems can be designed for the production of transportation fuels using inexpensive renewable sources of carbon. Ethanol and butanol are the chief alcohols in the transportation fuel which can be produced by the selected and optimized microbial consortia. Engineering fuel-producing microorganisms that secrete enzymes like cellulases and hemicellulases to break complex sugars before uptake and conversion into fuels may substantially reduce the production cost of fuel [65]. Similarly, robust-adaptive controlled devices can be designed and optimized for *in situ* delivery of therapeutics.

1.6 Challenges and Opportunities

Though engineered microorganisms have myriad ways that they can be applied for the synthesis of important molecules, there are many trade-offs that needs to be weighed, like:

Availability and cost starting materials

Selection of the optimum metabolic route and the corresponding genes encoding the enzymes for the production of the desired product

Selection of the appropriate microbial host

Stable and responsive genetic control elements that works in the selected host

Procedures to maximize yields, titres and productivity of the desired product

Quick fixtures or troubleshooting failed product formation at any step of development or production pipeline.

All the above design considerations are dependent on each other in the sense if the genes are not expressed at the set optimum, the enzyme coded by the gene will not function. Sophistication of the genetic tools available varies from host to host also processing conditions of growth; product separation and purification are not compatible with all hosts. These challenges may provide the opportunity for further developing robust and sensitive methods for the successful applications of metabolic engineering in a wide range of host for the production of economically important products. More so for the production of chemicals whose chemical synthesis is too complicated and can be achieved in higher living systems such as plants [69].

Future holds great promises for synthesizing tailor-made microorganism producing specific products from cheap starting materials. Such cell factories may be designed with pumps embedded in their membrane to pump out the final product out from the cells that reduces the purification costs of the desired product from the other thousand intermediate metabolites. Parts registry with all the updated and well-characterized parts should become one of the main sources for all the parts required to build the novel metabolic pathway. Software like RETROPATH [69] should be upgraded such that maximum yield can be predicted for a desired product from the chosen heterologous host. Computer-aided design of an enzyme that does not exist for a particular reaction would be an added advantage to design and create novel metabolic pathways [86]. Continued development of existing computer-aided tools alongside newer experimental methodologies can help garner the full potential of engineered microbes for the production of cost efficient natural and unnatural products.

References

1. Schwille P, Diez S (2009) Synthetic biology of minimal systems. Crit Rev Biochem Mol Biol 44(4):223–242
2. Porcar M, Danchin A, de Lorenzo V, Dos Santos VA, Krasnogor N, Rasmussen S, Moya A (2011) The ten grand challenges of synthetic life. Syst Synth Biol 5(1–2):1–9

3. Glass JI, Assad-Garcia N, Alperovich N, Yooseph S, Lewis MR, Maruf M, Hutchison CA, Smith HO, Venter JC (2006) Essential genes of a minimal bacterium. Proc Natl Acad Sci U S A 103(2):425–430
4. Lartigue C, Glass JI, Alperovich N, Pieper R, Parmar PP, Hutchison CA, Smith HO, Venter JC (2007) Genome transplantation in bacteria: changing one species to another. Science 317(5838):632–638
5. Gibson DG, Glass JI, Lartigue C, Noskov VN, Chuang RY, Algire MA, Benders GA, Montague MG, Ma L, Moodie MM, Merryman C (2010) Creation of a bacterial cell controlled by a chemically synthesized genome. Science 329(5987):52–56
6. McArthur GH, Fong SS (2009) Toward engineering synthetic microbial metabolism. BioMed Res Int 14:2010
7. Mushegian AR, Koonin EV (1996) A minimal gene set for cellular life derived by comparison of complete bacterial genomes. Proc Natl Acad Sci 93(19):10268–10273
8. Zhang LY, Chang SH, Wang J (2010) How to make a minimal genome for synthetic minimal cell. Protein Cell 1(5):427–434
9. Acevedo-Rocha CG, Fang G, Schmidt M, Ussery DW, Danchin A (2013) From essential to persistent genes: a functional approach to constructing synthetic life. Trends Genet 29(5):273–279
10. Salama NR, Shepherd B, Falkow S (2004) Global transposon mutagenesis and essential gene analysis of Helicobacter pylori. J Bacteriol 186(23):7926–7935
11. French CT, Lao P, Loraine AE, Matthews BT, Yu H, Dybvig K (2008) Large-scale transposon mutagenesis of Mycoplasma pulmonis. Mol Microbiol 69(1):67–76
12. Forsyth R, Haselbeck RJ, Ohlsen KL, Yamamoto RT, Xu H, Trawick JD, Wall D, Wang L, Brown-Driver V, Froelich JM, King P (2002) A genome-wide strategy for the identification of essential genes in Staphylococcus aureus. Mol Microbiol 43(6):1387–1400
13. Herring CD, Glasner JD, Blattner FR (2003) Gene replacement without selection: regulated suppression of amber mutations in Escherichia coli. Gene 311:153–163
14. Kobayashi K, Ehrlich SD, Albertini A, Amati G, Andersen KK, Arnaud M, Asai K, Ashikaga S, Aymerich S, Bessieres P, Boland F (2003) Essential Bacillus subtilis genes. Proc Natl Acad Sci 100(8):4678–4683
15. Fehér T, Papp B, Pál C, Pósfai G (2007) Systematic genome reductions: theoretical and experimental approaches. Chem Rev 107(8):3498–3513
16. Puchałka J, Oberhardt MA, Godinho M, Bielecka A, Regenhardt D, Timmis KN, Papin JA, dos Santos VA (2008) Genome-scale reconstruction and analysis of the Pseudomonas putida KT2440 metabolic network facilitates applications in biotechnology. PLoS Comput Biol 4(10):e1000210
17. Christian N, May P, Kempa S, Handorf T, Ebenhöh O (2009) An integrative approach towards completing genome-scale metabolic networks. Mol BioSyst 5(12):1889–1903
18. Zhang Y, Thiele I, Weekes D, Li Z, Jaroszewski L, Ginalski K, Deacon AM, Wooley J, Lesley SA, Wilson IA, Palsson B (2009) Three-dimensional structural view of the central metabolic network of Thermotoga maritima. Science 325(5947):1544–1549
19. Price ND, Reed JL, Palsson BØ (2004) Genome-scale models of microbial cells: evaluating the consequences of constraints. Nat Rev Microbiol 2(11):886–897
20. Holzhütter S, Holzhütter HG (2004) Computational design of reduced metabolic networks. Chembiochem 5(10):1401–1422
21. Brunk E, Neri M, Tavernelli I, Hatzimanikatis V, Rothlisberger U (2012) Integrating computational methods to retrofit enzymes to synthetic pathways. Biotechnol Bioeng 109:572–582
22. Carbonell P, Planson AG, Fichera D, Faulon JL (2011) A retrosynthetic biology approach to metabolic pathway design for therapeutic production. BMC Syst Biol 5:122
23. Cho A, Yun H, Park JHH, Lee SYY, Park S (2010) Prediction of novel synthetic pathways for the production of desired chemicals. BMC Syst Biol 4:35
24. Bachmann BO (2010) Biosynthesis: is it time to go retro? Nat Chem Biol 6:390–393
25. Cook A, Johnson P, Law J, Mirzazadeh M, Ravitz O, Simon A (2012) Computer-aided synthesis design: 40 years on. WIREs Comput Mol Sci 2:79–107
26. Edwards JS, Ibarra RU, Palsson BO (2001) *In silico* predictions of *Escherichia coli* metabolic capabilities are consistent with experimental data. Nat Biotechnol 19(2):125–130
27. Park JH, Lee KH, Kim TY, Lee SY (2007) Metabolic engineering of *Escherichia coli* for the production of L-valine based on transcriptome analysis and *in silico* gene knockout simulation. Proc Natl Acad Sci 104(19):7797–7802
28. Martin VJ, Pitera DJ, Withers ST, Newman JD, Keasling JD (2003) Engineering a mevalonate pathway in *Escherichia coli* for production of terpenoids. Nat Biotechnol 21(7):796–802
29. Kizer L, Pitera DJ, Pfleger BF, Keasling JD (2008) Application of functional genomics to pathway optimization for increased isoprenoid production. Appl Environ Microbiol 74(10):3229–3241
30. Alper H, Moxley J, Nevoigt E, Fink GR, Stephanopoulos G (2006) Engineering yeast transcription machinery for improved ethanol tolerance and production. Science 314(5805):1565–1568
31. Brochado AR, Matos C, Møller BL, Hansen J, Mortensen UH, Patil KR (2010) Improved vanillin production in baker's yeast through in silico design. Microb Cell Factories 9(1):1

32. Galdzicki M, Rodriguez C, Chandran D, Sauro HM, Gennari JH (2011) Standard biological parts knowledgebase. PLoS ONE 6(2):e17005
33. Medema MH, Breitling R, Bovenberg R, Takano E (2011) Exploiting plug-and-play synthetic biology for drug discovery and production in microorganisms. Nat Rev Microbiol 9(2):131–137
34. Heneghan MN, Yakasai AA, Halo LM, Song Z, Bailey AM, Simpson TJ, Cox RJ, Lazarus CM (2010) First heterologous reconstruction of a complete functional fungal biosynthetic multigene cluster. ChemBioChem 11(11):1508–1512
35. Hatzimanikatis V, Li C, Ionita JA, Henry CS, Jankowski MD, Broadbelt LJ (2005) Exploring the diversity of complex metabolic networks. Bioinformatics 21(8):1603–1609
36. Rodrigo G, Carrera J, Prather KJ, Jaramillo A (2008) DESHARKY: automatic design of metabolic pathways for optimal cell growth. Bioinformatics 24(21):2554–2556
37. Chou CH, Chang WC, Chiu CM, Huang CC, Huang HD (2009) FMM: a web server for metabolic pathway reconstruction and comparative analysis. Nucleic Acids Res 37(suppl 2):W129–W134
38. Pharkya P, Burgard AP, Maranas CD (2004) OptStrain: a computational framework for redesign of microbial production systems. Genome Res 14(11):2367–2376
39. Wang K, Neumann H, Peak-Chew SY, Chin JW (2007) Evolved orthogonal ribosomes enhance the efficiency of synthetic genetic code expansion. Nat Biotechnol 25(7):770–777
40. Mavromatis K, Chu K, Ivanova N, Hooper SD, Markowitz VM, Kyrpides NC (2009) Gene context analysis in the Integrated Microbial Genomes (IMG) data management system. PLoS ONE 4(11):e7979
41. Medema MH, Blin K, Cimermancic P, de Jager V, Zakrzewski P, Fischbach MA, Weber T, Takano E, Breitling R (2011) antiSMASH: rapid identification, annotation and analysis of secondary metabolite biosynthesis gene clusters in bacterial and fungal genome sequences. Nucleic Acids Res 39(suppl 2): W339–W346
42. Kanehisa M, Goto S, Kawashima S, Nakaya A (2002) The KEGG databases at GenomeNet. Nucleic Acids Res 30(1):42–46
43. Salis HM, Mirsky EA, Voigt CA (2009) Automated design of synthetic ribosome binding sites to control protein expression. Nat Biotechnol 27(10):946–950
44. Na D, Lee D (2010) RBSDesigner: software for designing synthetic ribosome binding sites that yields a desired level of protein expression. Bioinformatics 26(20):2633–2634
45. Villalobos A, Ness JE, Gustafsson C, Minshull J, Govindarajan S (2006) Gene designer: a synthetic biology tool for constructing artificial DNA segments. BMC Bioinformatics 7(1):285
46. Czar MJ, Cai Y, Peccoud J (2009) Writing DNA with GenoCAD™. Nucleic Acids Res 37(suppl 2): W40–W47
47. Hoover DM, Lubkowski J (2002) DNAWorks: an automated method for designing oligonucleotides for PCR-based gene synthesis. Nucleic Acids Res 30(10):e43
48. Bode M, Khor S, Ye H, Li MH, Ying JY (2009) TmPrime: fast, flexible oligonucleotide design software for gene synthesis. Nucleic Acids Res 37:W214–W221
49. Lee PA, Dymond JS, Scheifele LZ, Richardson SM, Foelber KJ, Boeke JD, Bader JS (2010) CLONEQC: lightweight sequence verification for synthetic biology. Nucleic Acids Res 38:2617–2623
50. Goler (2004) BioJADE: a design and simulation tool for synthetic biological systems. Master's thesis, MIT, MIT Computer Science and Artificial Intelligence Laboratory, May 2004
51. Flouris M, Bilas A (2004) Clotho: transparent data versioning at the block I/O level. In MSST:315–328
52. Rodrigo G, Carrera J, Jaramillo A (2007) Asmparts: assembly of biological model parts. Syst Synth Biol 1(4):167–170
53. Weeding E, Houle J, Kaznessis YN (2010) SynBioSS designer: a web-based tool for the automated generation of kinetic models for synthetic biological constructs. Brief Bioinform 11(4):394–402
54. Funahashi A, Morohashi M, Kitano H, Tanimura N (2003) Cell designer: a process diagram editor for gene-regulatory and biochemical networks. Biosilico 1(5):159–162
55. Becker SA, Feist AM, Mo ML, Hannum G, Palsson BØ, Herrgard MJ (2007) Quantitative prediction of cellular metabolism with constraint-based models: the COBRA Toolbox. Nat Protoc 2(3):727–738
56. Gevorgyan A, Bushell ME, Avignone-Rossa C, Kierzek AM (2011) SurreyFBA: a command line tool and graphics user interface for constraint-based modeling of genome-scale metabolic reaction networks. Bioinformatics 27(3):433–434
57. Le Fèvre F, Smidtas S, Combe C, Durot M, d'Alché-Buc F, Schachter V (2009) CycSim—an online tool for exploring and experimenting with genome-scale metabolic models. Bioinformatics 25(15):1987–1988
58. Cvijovic M, Olivares-Hernández R, Agren R, Dahr N, Vongsangnak W, Nookaew I, Patil KR, Nielsen J (2010) BioMet toolbox: genome-wide analysis of metabolism. Nucleic Acids Res 38(suppl 2): W144–W149
59. Yamada T, Letunic I, Okuda S, Kanehisa M, Bork P (2011) iPath2. 0: interactive pathway explorer. Nucleic Acids Res 39(suppl 2):W412–W415
60. Bates JT, Chivian D, Arkin AP (2011) GLAMM: genome-linked application for metabolic maps. Nucleic Acids Res 38:W400–W405
61. Wang HH, Isaacs FJ, Carr PA, Sun ZZ, Xu G, Forest CR, Church GM (2009) Programming cells by multiplex genome engineering and accelerated evolution. Nature 460(7257):894–898
62. Pósfai G, Plunkett G, Fehér T, Frisch D, Keil GM, Umenhoffer K, Kolisnychenko V, Stahl B, Sharma SS, De Arruda M, Burland V (2006) Emergent properties of reduced-genome Escherichia coli. Science 312(5776):1044–1046

63. Jensen PR, Hammer K (1998) The sequence of spacers between the consensus sequences modulates the strength of prokaryotic promoters. Appl Environ Microbiol 64(1):82–87
64. Smolke CD, Carrier TA, Keasling JD (2000) Coordinated, differential expression of two genes through directed mRNA cleavage and stabilization by secondary structures. Appl Environ Microbiol 66(12):5399–5405
65. Farmer WR, Liao JC (2000) Improving lycopene production in Escherichia coli by engineering metabolic control. Nat Biotechnol 18(5):533–537
66. Alper H, Stephanopoulos G (2007) Global transcription machinery engineering: a new approach for improving cellular phenotype. Metab Eng 9(3):258–267
67. Pfleger BF, Pitera DJ, Smolke CD, Keasling JD (2006) Combinatorial engineering of intergenic regions in operons tunes expression of multiple genes. Nat Biotechnol 24(8):1027–1032
68. Keasling JD (2010) Manufacturing molecules through metabolic engineering. Science 330(6009):1355–1358
69. Ro DK, Paradise EM, Ouellet M, Fisher KJ, Newman KL, Ndungu JM, Ho KA, Eachus RA, Ham TS, Kirby J, Chang MC (2006) Production of the antimalarial drug precursor artemisinic acid in engineered yeast. Nature 440(7086):940–943
70. Chang MC, Eachus RA, Trieu W, Ro DK, Keasling JD (2007) Engineering Escherichia coli for production of functionalized terpenoids using plant P450s. Nat Chem Biol 3(5):274–277
71. Dietrich JA, Yoshikuni Y, Fisher KJ, Woolard FX, Ockey D, McPhee DJ, Renninger NS, Chang MC, Baker D, Keasling JD (2009) A novel semi-biosynthetic route for artemisinin production using engineered substrate-promiscuous P450BM3. ACS Chem Biol 4(4):261–267
72. Ajikumar PK, Xiao WH, Tyo KE, Wang Y, Simeon F, Leonard E, Mucha O, Phon TH, Pfeifer B, Stephanopoulos G (2010) Isoprenoid pathway optimization for Taxol precursor overproduction in Escherichia coli. Science 330(6000):70–74
73. Müller U, Van Assema F, Gunsior M, Orf S, Kremer S, Schipper D, Wagemans A, Townsend CA, Sonke T, Bovenberg R, Wubbolts M (2006) Metabolic engineering of the E. colil-phenylalanine pathway for the production of d-phenylglycine (d-Phg). Metab Eng 8(3):196–208
74. Karlsson M, Weber W (2012) Therapeutic synthetic gene networks. Curr Opin Biotechnol. doi:10.1016/j.copbio.2012.1001.1003
75. Ruder WC, Lu T, Collins JJ (2011) Synthetic biology moving into the clinic. Science 333:1248–1252
76. Weber W, Fussenegger M (2012) Emerging biomedical applications of synthetic biology. Nat Rev Genet 13:21–35
77. Ye H, Daoud-El Baba M, Peng RW, Fussenegger M (2011) A synthetic optogenetic transcription device enhances blood-glucose homeostasis in mice. Science 332:1565–1568
78. Xie Z, Wroblewska L, Prochazka L, Weiss R, Benenson Y (2011) Multiinput RNAi-based logic circuit for identification of specific cancer cells. Science 333:1307–1311
79. Kemmer C, Gitzinger M, Daoud-El Baba M, Djonov V, Stelling J, Fussenegger M (2010) Self-sufficient control of urate homeostasis in mice by a synthetic circuit. Nat Biotechnol 28:355–360
80. Hanai T, Atsumi S, Liao JC (2007) Engineered synthetic pathway for isopropanol production in Escherichia coli. Appl Environ Microbiol 73(24):7814–7818
81. Atsumi S, Hanai T, Liao JC (2008) Non-fermentative pathways for synthesis of branched-chain higher alcohols as biofuels. Nature 451(7174):86–89
82. Steen EJ, Kang Y, Bokinsky G, Hu Z, Schirmer A, McClure A, Del Cardayre SB, Keasling JD (2010) Microbial production of fatty-acid-derived fuels and chemicals from plant biomass. Nature 463(7280):559–562
83. Bayer TS, Widmaier DM, Temme K, Mirsky EA, Santi DV, Voigt CA (2009) Synthesis of methyl halides from biomass using engineered microbes. J Am Chem Soc 131(18):6508–6515
84. Dunlop MJ, Dossani ZY, Szmidt HL, Chu HC, Lee TS, Keasling JD, Hadi MZ, Mukhopadhyay A (2011) Engineering microbial biofuel tolerance and export using efflux pumps. Mol Syst Biol 1:7(1)
85. Prather KL, Martin CH (2008) De novo biosynthetic pathways: rational design of microbial chemical factories. Curr Opin Biotechnol 19(5):468–474
86. Siegel JB, Zanghellini A, Lovick HM, Kiss G, Lambert AR, Clair JL, Gallaher JL, Hilvert D, Gelb MH, Stoddard BL, Houk KN (2010) Computational design of an enzyme catalyst for a stereoselective bimolecular Diels-Alder reaction. Science 329(5989):309–313

Computational Proteomics

2

Debasree Sarkar and Sudipto Saha

2.1 Introduction

Proteomics is the large-scale study of proteins, particularly their structures and functions, and it is the leading area of research in biological science in the twenty-first century. Proteomics represents the effort to establish the identities, quantities, structures, and biochemical and cellular functions of all proteins in an organism, organ, or organelle. In addition, proteomics also describes how these properties vary in space, time, or physiological state. The term *proteomics* was first coined in 1997 to make an analogy with genomics, the study of the genome. The proteome denotes the total complement of proteins found in a complete genome or a specific tissue [1]. The traditional approach of studying the functions of proteins is to consider one or two proteins at a time using biochemical characterization and genetic methods. Due the advent of high-throughput approaches including 2D gel electrophoresis and mass spectrometry (MS)-based proteomics, we can study thousands of proteins in a single experiment [2]. Since high-throughput proteomics generates huge amount of data, these may be prone to false positive identifications. Hence, it is essential to be cautious while interpreting such results/data. To overcome it, statistical and computational tools are used to gain confidence in interpreting the result. The workflow of proteomics includes protein fractionation using 1D/2D electrophoresis followed by protein identification by MS. 2D separation is based on size and charge, where the first step is to separate the complex mixture of proteins based on charge or isoelectric point, called isoelectric focusing and then separate based on size (SDS-PAGE). After gel separation, proteins are excised and digested by enzyme trypsin/chymotrypsin into many peptides, which have specific cutting sites in the primary amino acid sequences. These peptides are subjected to mass spectrometry for identification based on mass by charge (m/z) ratio. MS can be grouped into two classes based on ionization process, matrix-assisted laser desorption ionization (MALDI) and electrospray ionization (ESI). The Nobel Prize in Chemistry 2002 was awarded to Koichi Tanaka for the development of soft desorption ionization methods for mass spectrometric analyses of biological macromolecules. MS-based proteomics can be implemented using top-down approach involving MS of whole protein ions and bottom-up approach, where peptides are subjected to MS and eventually proteins are predicted/inferred based on peptide identification as shown in Fig. 2.1. Due to instrument constraint, bottom-up approach is more popular in biomedical research.

D. Sarkar • S. Saha (✉)
Centre of Excellence in Bioinformatics,
Bose Institute, Kolkata, India
e-mail: ssaha4@gmail.com

S. Singh (ed.), *Systems Biology Application in Synthetic Biology*,
DOI 10.1007/978-81-322-2809-7_2

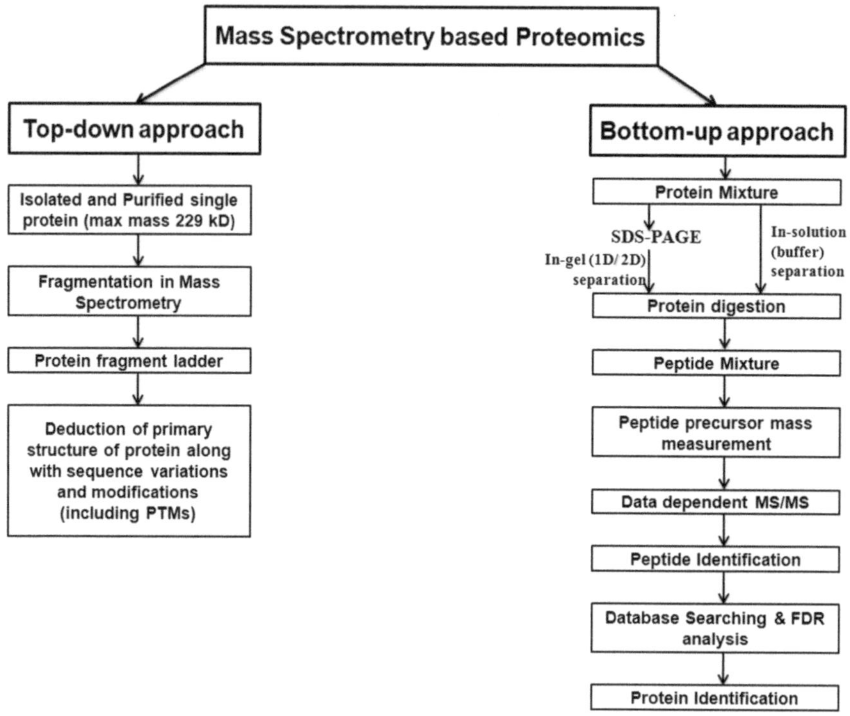

Fig. 2.1 Workflow for mass spectrometry-based proteomics employed in biomedical research

For complex mixtures like plasma proteins from blood, the peptide mixtures are separated by liquid chromatography and then subjected to mass spectrometry. Each peptide precursor is further fragmented to y and b ions for sequence order, which is termed as tandem MS or MS/MS. Finally the peptides are identified and proteins are predicted by sequence database matching. However, in the absence of genomic DNA, cDNAs, ESTs, or protein sequences for a specific organism, the identification of peptides from MS/MS spectra can be done by a database-independent approach which is termed as de novo sequencing.

In proteomics, many computational tools and software are required for which a pipeline is necessary for quality control. These include the preprocessing of MS spectra, protein identification using search engines, quantitation of protein, and finally storage of the MS data. For preprocessing step, deconvolution, intensity normalization, and filtration of low-quality spectra are required. Deconvolution is an application of a mathematical algorithm to transform raw data into a meaningful format for further analysis, involving background subtraction, noise removal, charge state deconvolution, and deisotoping. Normalization techniques commonly used include normalization to base peak, rank-based normalization, and local normalization to highest intensity in a user-defined m/z bin size. The protein identification and characterization is done by database searching of MS/MS data [3]. The search engines commonly used are Mascot [4], Sequest [5], and X!Tandem [6]. All the search engines require additional information in the form of search parameters including name of the sequence database, taxonomy, mass tolerance, enzyme (trypsin most commonly used), and posttranslational modifications. There is a challenge in protein inference from peptide sequences in shotgun proteomics, where proteins from a cell lysate are digested to peptides. In addition, there is a bigger challenge in protein quantification from complex peptide mixture including plasma samples. The popular software tools for measuring protein abundance are Scaffold [7] and Rosetta Elucidator [8], which use spectral count and peptide intensity, respectively.

Table 2.1 Useful programs for data analysis of MS-based proteomics

Preprocessing of MS spectra	Search engines	Quantification	Repository
Mass-Up	Mascot	Scaffold	PRIDE
mMass	Sequest	Elucidator	Tranche
AMDIS	X!Tandem	Census	GPMDB
Ms-Deconv	OMSSA	MaxQuant	PeptideAtlas
Abacus	MassMatrix	XPRESS	CPAS

There are MS data repositories allowing data submission and retrieval for collaborative and public users. The commonly useful programs for MS-based data analyses are listed in Table 2.1.

2.2 Protein Identification

Protein identification relies on peptide MS/MS spectra matching to the protein sequence database. The selection of search engine and right database is an important step for identification of proteins. Many a times the same peptide sequence can be present in multiple different proteins or protein isoforms; thus in such cases it is difficult to assign a peptide to a protein [9]. In shotgun proteomics, the standard criterion for inferring protein is to identify at least two unique peptides and with reasonable amino acid sequence coverage. The selection of identified peptides from spectra is based on scores above a threshold value. Different scoring schemes have been developed for peptide matching. For example, Mascot [4] and OMSSA [10] use probability-based scoring, while Sequest [5] uses descriptive approach. For large-scale studies of complex mixture of proteins, the False Discovery Rate (FDR) is used for peptide selection. All the search engines require additional information in the form of search parameters. The critical parameters are discussed below.

2.2.1 Sequence Database

In shotgun proteomics approach, the connectivity between peptides and proteins is lost in the enzymatic digestion stage. The task of assembling the protein sequences from identified peptides is done by searching in sequence database using computational tools, which requires selection of a reference protein sequence database. The most commonly used databases are UniProt/Swiss-Prot and RefSeq from NCBI. Both of these databases are non-redundant and well curated and thus help in biological data interpretation. In case an organism is not well represented in protein databases, EST databases are used.

2.2.2 Taxonomy

The protein sequence databases contain taxonomy information, and most search engines allow users to restrict the search to entries for a particular organism or taxonomic rank. Limiting the taxonomy makes the database smaller and removes the homologous proteins from other species. This eventually speeds up the search process and avoids misleading matches. However, when searching proteins for poorly represented species in the databases, it is better to specify higher-order taxonomy. The size of the database in terms of the number of proteins has an effect in the search result and protein scores.

2.2.3 Enzyme

The cleavage method needs to be selected in the search form. The most widely used enzyme is trypsin, which cleaves after arginine and lysine if they are not followed by proline. In practice, the cleavage methods are not 100 % specific and thus the search form allows users to specify the missed cleavages of one or maybe two.

2.2.4 Modifications

There are two types of modifications that need to be specified in the database searching. First, fixed modifications correspond to mass change of an amino acid and do not take a longer search time, for example, alkylation of cysteine, where all cysteines are modified and there is change in the mass of cysteine. Second, variable modifications, in which, the modifications do not apply to all the instances of a residue. For example, not all serines in a peptide are phosphorylated. This type of search increases the time taken for a search since the software considers all the possible arrangements of modified and unmodified residues that fit to the peptide molecular mass.

2.2.5 Peak List File Format

There are a number of different file formats for peak lists. Mascot uses MGF (Mascot Generic Format), whereas Sequest supports DTA and PKA formats. mzML is the standard interchange format supported by proteomics standard initiatives, which can be used for raw and peak lists.

2.2.6 Mass Tolerance

Most search engines support peptide mass tolerance for precursors and fragments. The peptide tolerance in narrow windows of 1 and 2 Da is preferred. Specifying less than 1 mass tolerance may lose the sensitivity of the match.

2.2.7 False Discovery Rate (FDR)

Many search engines and scoring systems provide an option of statistical validation of the results and use a decoy database to estimate FDR. A decoy database is a database of amino acid sequences that is derived from the original protein database (called the target database) by reversing the target sequences, shuffling the target sequences, or generating the decoy sequences at random. Generally FDR is calculated on peptide hits and a threshold cutoff value of 1 % is allowed.

2.3 Quantitative Proteomics

Quantitative proteomics deals in relative protein expression levels between two or more different pools of proteins. It is used to detect the difference in protein expression profiles among tissues, cell cultures, or organisms. Most commonly, it is used to compare expression profiles between a healthy cell and a diseased cell. The data comparison with diseased cells/tissues can be used for biomarker or drug discovery. 2D gel-based proteomics and difference gel electrophoresis (DIGE), which uses fluorescence-based labeling of the proteins prior to separation, are current approaches for the 2-DE-based study of proteomes [11]. Recently, shotgun proteomics approaches are being used for protein expression profiling in two different ways: (1) label-free method and (2) stable isotope labeling methods. In addition to assembling peptides to proteins, quantitative proteomics data deals with protein abundance ratios.

2.3.1 Label-Free Quantification Methods

In label-free quantification approach, relative abundance of peptides in two or more biological samples is determined, based either on spectral counting or on precursor ion signal intensity. Many automated software tools including scaffold use spectral count as a quantitative value for protein abundance. Spectral count is the number of peptides identified from a protein in each sample. Peptide fragment ion intensities are used by Rosetta Elucidator, which measures and compares the signal intensities of peptide precursor ions. Biological samples have a wide range of protein abundance values, and mass spectrometers are not well equipped to detect a dynamic range. For example, blood samples contain a few thousands of proteins including tissue leakage proteins and cytokines in low abundance. The peptides from highly abundant proteins often mask the low-abundant proteins. The spectra or the intensity profiling methods compare the peak intensities across different LC-MS runs, and it is required to perform replicate measurements to estimate the variance.

2.3.1.1 Statistical Analysis

Quantitative proteomics deals with comparing protein abundance values in two different conditions and across replicated experiments. Data normalization is essential for the comparison of the LC-MS intensity/spectral profiles.

Normalized spectral abundance factor (NSAF) [12], Z-score [13], and a few other scoring systems are used to perform the normalization step. After normalization, fold change and testing of significance using *t*-test (similar to microarray studies) are carried out. A volcano plot helps to understand the level of significance and magnitude of changes observed in a quantitative proteomics study. The fold change on the log2 scale is placed on the horizontal axis and the p-value on the -log10 scale is placed on the vertical axis [14] , as shown in Fig. 2.2.

2.3.1.2 Visualization and Pathway Analysis

Heat map and clustering analysis allows visualization and interpretation of the expression data. For further interpretation, the expression data set can be uploaded in pathway analysis software tools like Ingenuity Pathway Analysis (IPA) [15] and Pathway Studio [16] for identification of significant pathways that have changed in different conditions. IPA is a web-based software application for the analysis, integration, and interpretation of proteomics data, in which, the back-end data has been manually curated. Pathway Studio also enables the analysis and visualization of proteomics expression and pathway curation, but here the back-end data has been collected by text mining.

2.3.2 Applications of Quantitative Proteomics

A mapping of human proteome of adult tissues, fetal tissues, and hematopoietic stem cells (HSCs) was performed using shotgun LC MS/MS. Developmental stage-specific differential expression of protein complexes in fetal and adult liver tissues was identified. This resulted in large human proteome catalog of 17,294 genes [17].

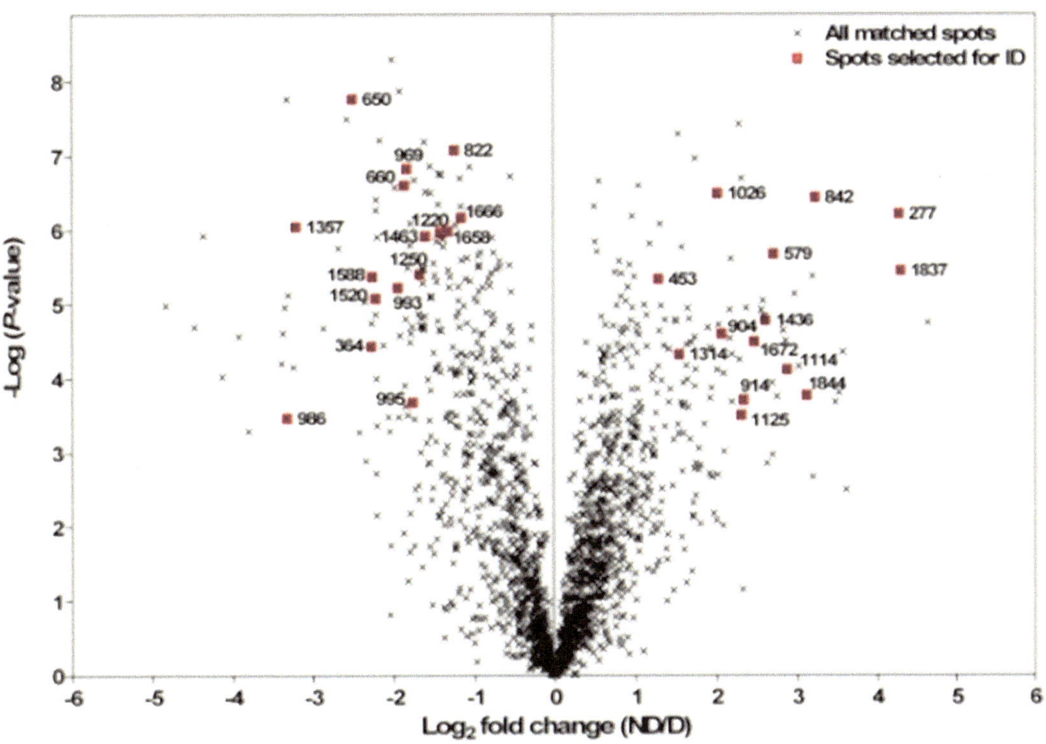

Fig. 2.2 Volcano plot for graphical representation of quantitative proteomics data

The protein composition may be associated with disease processes in the organism and thus have potential utility as diagnostic markers. Proteins are closer to the actual disease process, in most cases, than parent genes. Proteins are ultimate regulators of cellular function. Most cancer biomarkers are proteins, e.g., detection of PSA is a surrogate for early detection of prostate cancer. Large screening trials have shown that PSA nearly doubles the rate of detection when combined with other methods. Based on these data, PSA testing was approved by the US FDA for the screening and early detection of prostate cancer.

2.4 Interaction Proteomics

Proteins interact with each other to form functional units like networks and pathways. Individual protein functions can be revealed through participation in specific interaction networks. The two commonly used techniques to study protein-protein interactions (PPIs) are yeast 2-hybrid (Y2H) and affinity purification-mass spectrometry (AP-MS). Yeast and human PPIs have been extensively studied using these two methods. The former deals with binary interactions and later identifies multi-protein complexes. The bait protein is the protein of interest while the prey proteins are the proteins associated with the bait protein. Both the methods are incomplete and the network is dependent on the technology (Fig. 2.3). AP-MS combines the specificity of antibody-based protein purification with the sensitivity of mass spectrometry to identify and quantify putative interacting proteins. There are key issues in both the technologies. In Y2H, if a protein interacts in the presence of two or more proteins, such instances cannot be captured (Fig. 2.3b). For example, active PP2A holoenzyme requires the catalytic, regulatory, and structural units to form a complex. Such studies are possible only by AP-MS. However, AP-MS has its own limitations. First, there is variability

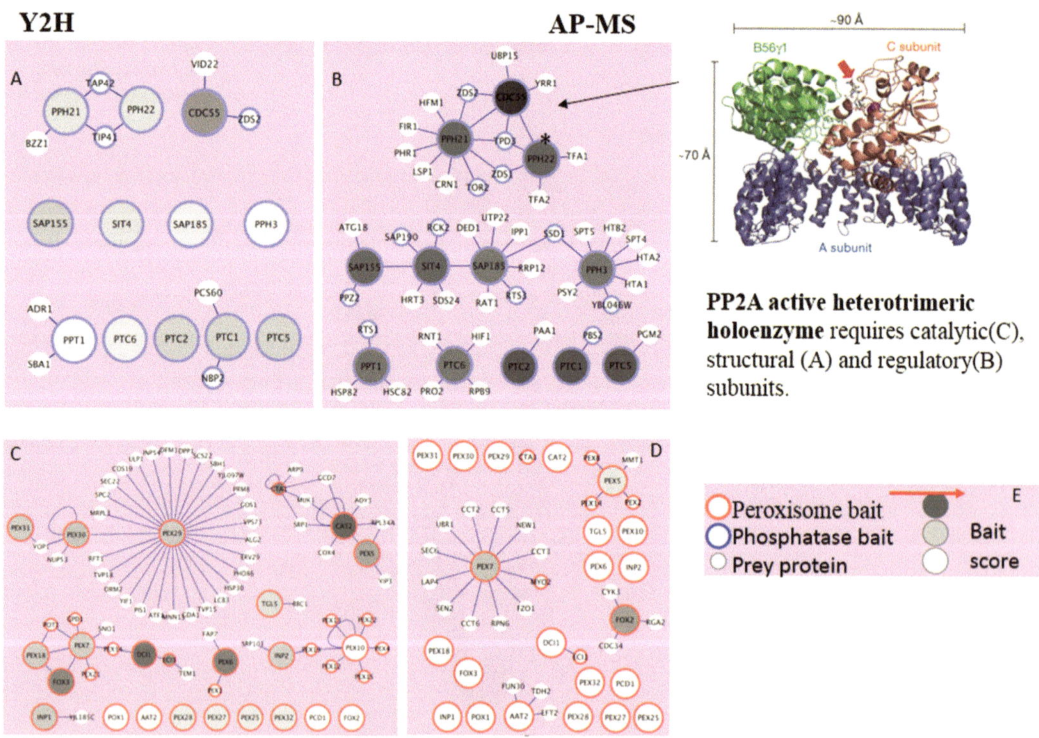

Fig. 2.3 Comparative analysis of selected Y2H and AP-MS yeast networks (Adapted from Saha et al. [18])

in AP-MS replicated experiments in terms of prey proteins identified. Second, there may be many nonspecific binders. Third, not all bait proteins are expressed well in the transfected cells, and it is difficult to identify them if it is expressed in small vesicles like peroxisomes (Fig. 2.3d). To overcome these problems, several statistical tools have been developed [18].

2.4.1 Scoring Systems for PPIs

The analysis of protein interaction networks and protein complexes are very important for understanding the cellular process. Development of computational tools for identifying true interactors and modeling bait-prey and prey-prey interactions is a rapidly growing field of research. Socio-affinity score was first used in yeast interactome study using AP-MS [19]. The major drawback of this method was that all the prey proteins have to be used as bait again for applying this score. The other scores used are NSAF [20] and ROCS [21]. CompPASS (Comparative Proteomics Analysis Software Suite) uses D-score and is designed to help facilitate the identification of high confidence candidate interacting proteins from IP-MS/MS data [22]. CRAPome [23] is a repository of AP-MS background contaminant data for human and yeast and includes computational tools like SAINT [24] and SAINTexpress [25] for AP-MS data analysis.

2.4.2 PPI Databases

There are some comprehensive highly curated databases for storing information about PPIs and protein complexes [26]. Some of them are organism specific, like the Human Protein Reference Database (HPRD) [27] and Comprehensive Resource of Mammalian protein complexes (CORUM) [28], while some do not restrict to species like IntAct [29], DIP [30], and BioGRID [31]. The STRING database [32] provides predicted as well as manually curated PPIs of a wide range of species.

2.4.3 Applications of Interaction Proteomics

Interaction proteomics includes physical PPI networks and the protein complexes formed by biochemical events to serve a distinct biological function as a complex. The protein interactome describes the full repertoire of PPIs within a biological system. Recently the BioPlex (biophysical interactions of ORFEOME-derived complexes) network [33] was generated from thousands of human cell lines each expressing a tagged version of a protein from the human ORFEOME collection [34]. AP-MS-based method was used as the building blocks of this network. Other interesting networks developed from the same group are the human autophagy interaction network (AIN), the human interaction network for ER-associated degradation (INfERAD), and the mitochondrial networks.

2.5 Metaproteomics

The environmental metaproteomic measurements for many different microbes including uncultured organisms in mixed communities can be studied by using MS-based proteomics and computational tools for characterization of complete proteins expressed by microbial community in an environmental sample [35, 36]. A variety of research areas including bioremediation, bioenergy, and human health can be addressed using metaproteomics. The characterization of microbial species and their impact on the human gut in healthy and disease patients can have profound implications on human health. Some useful computational tools used in metaproteomics analyses are Unipept [37], MetaProteomeAnalyzer (MPA) [38], and Pipasic [39].

2.6 Proteomics Standard Initiative

The Proteomics Standards Initiative (PSI) aims to define community standards for data representation in proteomics and to facilitate data comparison, exchange, and verification. The PSI from the Human Proteome Organization (HUPO-PSI) has defined standards for proteomics data representation as well as guidelines that state the minimum information that should be included when reporting a proteomics experiment (MIAPE) [40]. Such minimum information must describe the complete experiment, including both experimental protocols and data processing methods, allowing a critical evaluation of the whole process and the potential recreation of the work. For interaction proteomics, the PSI-MI interchange format [41] was developed which contains controlled vocabularies designed by a consortium of molecular interaction database providers including BioGRID, DIP, IntAct, and HPRD. PSI-MI integrates with Biological Pathway Exchange (BioPAX), which is the standard language to represent biological pathways [42]. BioPAX and PSI-MI are designed for data exchange from databases as well as pathway and network data integration. Tools are available for converting PSI-MI format to BioPAX.

2.7 Data Repositories

Proteomics studies generate large volumes of raw experimental data. Hence, to facilitate the dissemination of these data, centralized data repositories were developed that make the data and results accessible to proteomics researchers and biologists [43]. PRIDE, the "Proteomics Identifications database," is a public repository of protein and peptide identifications for the proteomics community [44]. It focuses mainly on shotgun mass spectrometry proteomics data, and proteomics researchers can deposit their MS/MS proteomics data sets according to the guidelines of the ProteomeXchange (PX) consortium. Since PRIDE is a web application, submission, searching, and data retrieval can all be performed using an Internet browser. PRIDE allows users to search by experiment accession number, protein accession number, literature reference, and sample parameters including species, tissue, subcellular location, and disease state. Data can be retrieved either as machine-readable PRIDE/mzData XML files (the latter for mass spectra without identifications), or as human-readable HTML files. Tranche [45] is another distributed data repository designed to redundantly store and disseminate data sets for the proteomics community. Other repositories such as PRIDE, PeptideAtlas, and Human Proteinpedia interact with Tranche as the preferred mechanism for storing and disseminating large MS data files.

References

1. Colinge J, Bennett KL (2007) Introduction to computational proteomics. PLoS Comput Biol 3(7):e114
2. Nilsson T, Mann M, Aebersold R et al (2010) Mass spectrometry in high-throughput proteomics: ready for the big time. Nat Methods 7(9):681–685. doi:10.1038/nmeth0910-681
3. Cottrell JS (2011) Protein identification using MS/MS data. J Proteomics 74(10):1842–1851. doi:10.1016/j.jprot.2011.05.014
4. Perkins DN, Pappin DJ, Creasy DM et al (1999) Probability-based protein identification by searching sequence databases using mass spectrometry data. Electrophoresis 20(18):3551–3567
5. Eng JK, McCormack AL, Yates JR (1994) An approach to correlate tandem mass spectral data of peptides with amino acid sequences in a protein database. J Am Soc Mass Spectrom 5(11):976–989. doi:10.1016/1044-0305(94)80016-2
6. Fenyö D, Beavis RC (2003) A method for assessing the statistical significance of mass spectrometry-based protein identifications using general scoring schemes. Anal Chem 75(4):768–774
7. Searle BC (2010) Scaffold: a bioinformatic tool for validating MS/MS-based proteomic studies. Proteomics 10(6):1265–1269. doi:10.1002/pmic.200900437
8. Neubert H, Bonnert TP, Rumpel K et al (2008) Label-free detection of differential protein expression by LC/MALDI mass spectrometry. J Proteome Res 7(6):2270–2279. doi:10.1021/pr700705u
9. Nesvizhskii AI, Aebersold R (2005) Interpretation of shotgun proteomic data: the protein inference problem. Mol Cell Proteomics 4(10):1419–1440
10. Geer LY, Markey SP, Kowalak JA (2004) Open mass spectrometry search algorithm. J Proteome Res 3(5):958–964
11. Rabilloud T, Lelong C (2011) Two-dimensional gel electrophoresis in proteomics: a tutorial. J Proteomics 74(10):1829–1841. doi:10.1016/j.jprot.2011.05.040

12. Paoletti AC, Parmely TJ, Tomomori-Sato C (2006) Quantitative proteomic analysis of distinct mammalian mediator complexes using normalized spectral abundance factors. Proc Natl Acad Sci 103(50):18928–18933
13. Lu P, Vogel C, Wang R (2007) Absolute protein expression profiling estimates the relative contributions of transcriptional and translational regulation. Nat Biotechnol 25(1):117–124
14. Ntai I, Kim K, Fellers RT et al (2014) Applying label-free quantitation to top down proteomics. Anal Chem 86(10):4961–4968. doi:10.1021/ac500395k
15. Müller T, Schrötter A, Loosse C et al (2011) Sense and nonsense of pathway analysis software in proteomics. J Proteome Res 10(12):5398–5408. doi:10.1021/pr200654k
16. Nikitin A, Egorov S, Daraselia N et al (2003) Pathway studio – the analysis and navigation of molecular networks. Bioinformatics 19(16):2155–2157
17. Kim MS, Pinto SM, Getnet D et al (2014) A draft map of the human proteome. Nature 509(7502):575–581. doi:10.1038/nature13302
18. Saha S, Kaur P, Ewing RM (2010) The bait compatibility index: computational bait selection for interaction proteomics experiments. J Proteome Res 9(10):4972–4981. doi:10.1021/pr100267t
19. Gavin AC, Aloy P, Grandi P et al (2006) Proteome survey reveals modularity of the yeast cell machinery. Nature 440(7084):631–636
20. Sardiu ME, Cai Y, Jin J (2008) Probabilistic assembly of human protein interaction networks from label-free quantitative proteomics. Proc Natl Acad Sci 105(5):1454–1459. doi:10.1073/pnas.0706983105
21. Dazard JE, Saha S, Ewing RM (2012) ROCS: a reproducibility index and confidence score for interaction proteomics studies. BMC Bioinforma 13:128. doi:10.1186/1471-2105-13-128
22. Sowa ME, Bennett EJ, Gygi SP et al (2009) Defining the human deubiquitinating enzyme interaction landscape. Cell 138(2):389–403. doi:10.1016/j.cell.2009.04.042
23. Mellacheruvu D, Wright Z, Couzens AL et al (2013) The CRAPome: a contaminant repository for affinity purification-mass spectrometry data. Nat Methods 10(8):730–736. doi:10.1038/nmeth.2557
24. Choi H, Larsen B, Lin ZY et al (2011) SAINT: probabilistic scoring of affinity purification-mass spectrometry data. Nat Methods 8(1):70–73. doi:10.1038/nmeth.1541
25. Teo G, Liu G, Zhang J et al (2014) SAINTexpress: improvements and additional features in Significance Analysis of INTeractome software. J Proteomics 100:37–43. doi:10.1016/j.jprot.2013.10.023
26. Mathivanan S, Periaswamy B, Gandhi TK et al (2006) An evaluation of human protein-protein interaction data in the public domain. BMC Bioinforma 7(5):S19
27. Goel R, Muthusamy B, Pandey A et al (2011) Human protein reference database and human proteinpedia as discovery resources for molecular biotechnology. Mol Biotechnol 48(1):87–95. doi:10.1007/s12033-010-9336-8
28. Ruepp A, Waegele B, Lechner M et al (2010) CORUM: the comprehensive resource of mammalian protein complexes – 2009. Nucleic Acids Res 38(Database issue):D497–D501. doi:10.1093/nar/gkp914
29. Orchard S, Ammari M, Aranda B et al (2014) The MIntAct project – IntAct as a common curation platform for 11 molecular interaction databases. Nucleic Acids Res 42(Database issue):D358–D363. doi:10.1093/nar/gkt1115
30. Salwinski L, Miller CS, Smith AJ et al (2004) The database of interacting proteins: 2004 update. Nucleic Acids Res 32(Database issue):D449–D451
31. Oughtred R, Chatr-Aryamontri A, Breitkreutz BJ (2016) BioGRID: a resource for studying biological interactions in yeast. Cold Spring Harb Protoc 2016(1):pdb.top080754. doi:10.1101/pdb.top080754
32. Szklarczyk D, Franceschini A, Wyder S et al (2015) STRING v10: protein-protein interaction networks, integrated over the tree of life. Nucleic Acids Res 43(Database issue):D447–D452. doi:10.1093/nar/gku1003
33. Huttlin EL, Ting L, Bruckner RJ et al (2015) The BioPlex network: a systematic exploration of the human interactome. Cell 162:425–440
34. Yang X, Boehm JS, Yang X et al (2011) A public genome-scale lentiviral expression library of human ORFs. Nat Methods 8(8):659–661. doi:10.1038/nmeth.1638
35. Hettich RL, Pan C, Chourey K et al (2013) Metaproteomics: harnessing the power of high performance mass spectrometry to identify the suite of proteins that control metabolic activities in microbial communities. Anal Chem 85(9):4203–4214. doi:10.1021/ac303053e
36. Abraham PE, Giannone RJ, Xiong W et al (2014) Metaproteomics: extracting and mining proteome information to characterize metabolic activities in microbial communities. Curr Protoc Bioinformatics 46:13.26:13.26.1–13.26.14
37. Mesuere B, Debyser G, Aerts M et al (2015) The Unipept metaproteomics analysis pipeline. Proteomics 15(8):1437–1442. doi:10.1002/pmic.201400361
38. Muth T, Behne A, Heyer R et al (2015) The MetaProteomeAnalyzer: a powerful open-source software suite for metaproteomics data analysis and interpretation. J Proteome Res 14(3):1557–1565. doi:10.1021/pr501246w
39. Penzlin A, Lindner MS, Doellinger J et al (2014) Pipasic: similarity and expression correction for strain-level identification and quantification in metaproteomics. Bioinformatics 30(12):i149–i156. doi:10.1093/bioinformatics/btu267

40. Taylor CF, Paton NW, Lilley KS et al (2007) The minimum information about a proteomics experiment (MIAPE). Biotechnology 25(8):887–893
41. Hermjakob H, Montecchi-Palazzi L, Bader G et al (2004) The HUPO PSI's molecular interaction format – a community standard for the representation of protein interaction data. Nat Biotechnol 22(2):177–183
42. Demir E, Cary MP, Paley S et al (2010) The BioPAX community standard for pathway data sharing. Nat Biotechnol 28(9):935–942. doi:10.1038/nbt.1666
43. Riffle M, Eng JK (2009) Proteomics data repositories. Proteomics 9(20):4653–4663. doi:10.1002/pmic.200900216
44. Vizcaíno JA, Côté RG, Csordas A et al (2013) The PRoteomics IDEntifications (PRIDE) database and associated tools: status in 2013. Nucleic Acids Res 41(Database issue):D1063–D1069. doi:10.1093/nar/gks1262
45. Smith BE, Hill JA, Gjukich MA (2011) Tranche distributed repository and ProteomeCommons.org. Methods Mol Biol 696:123–145

3 Design, Principles, Network Architecture and Their Analysis Strategies as Applied to Biological Systems

Ahmad Abu Turab Naqvi and Md. Imtaiyaz Hassan

3.1 Introduction

Life at the level of a unit seems so simple and concrete. But, going deeper into the modular level of its formation, looks more complex and abstract than the painting of an abstractionist. Man, from the early days of the evolution of his consciousness, is constantly working by constructing all possible ways to observe and unfold the complexity of life, in terms of biology. The differentiation of all life forms is achievable in terms of Systems Theory [37]. By considering distinct modules of life as system, we can go further into the systems bringing forward a better understanding of its organisation. In terms of biology, a community of living organisms, a population, an individual organism and going more further down to the cellular level a single cell can be characterized as a system [37].

Systems biology facilitates the observation of biological systems at the molecular level to understand the underlying dynamics [19]. Foundations of Systems Biology as most of the Systems biologists agree with date back to the year 1948 in the works of Cybernetics carried out by Norbert Weiner [39]. However the word "systems biology" came into use during 1960s [33]. Though, various approaches have been assigned to understand the interior mechanism of the biological systems in the past, most of the studies were based on obtaining the physiological level of understanding rather than that of the molecular level. The factors behind this limitation in approach are the inability to make microscopic observations during that time. Though countless attempts have been made to explore biological systems to understand their working mechanisms, these approaches were limited due to lack of information about these systems at the molecular level. With the advancement in molecular biology after the Watson and Crick's discovery of the structure of DNA [38], the field of systems biology has been growing [19, 40]. Currently, while exploring the mechanism of complex biological systems, focus is laid on the molecular framework of the systems with respect to its underlying biological components such as genes, proteins and other macromolecular species. In this chapter, we will try to discuss established facts about biological networks, the inbuilt design principles embedded in these networks and some analysis strategies applied to these systems for dynamical observation of biological systems.

A.A.T. Naqvi
Department of Computer Science, Jamia Millia Islamia, Jamia Nagar, New Delhi 110025, India

Md.I. Hassan (✉)
Center for Interdisciplinary Research in Basic Sciences, Jamia Millia Islamia, Jamia Nagar, New Delhi 110025, India
e-mail: mihassan@jmi.ac.in

S. Singh (ed.), *Systems Biology Application in Synthetic Biology*,
DOI 10.1007/978-81-322-2809-7_3

3.2 Biological Networks – Architecture and Design Principle

Biological networks can be defined in terms of the Graph Theory [4]. Graphs are mathematical structures that are used to model pairwise relationships between the objects. Biological networks similarly are a collection of molecular species (nodes) which have interconnections (edges). The hierarchy of biological networks therefore depends on its components. Networks can be either simple or complex in nature. The present chapter address the architecture of such networks and their design principles to understand the working mechanisms of various biological networks. For the brevity of the subject and taking the importance of the subject into consideration, we have confined the description of biological networks at the molecular level only. Cell, being a subtle example of biological systems, encapsulates most intricate forms of biological networks inside its boundaries. Considering the complexity of biological systems, it would be an abstract idea to consider biological networks as distinct units. Each of the network inside the cell is associated with several other networks that function parallelly and modulate the cellular activity. Biological activities therefore are a consequence of the orchestra of functioning of the many biological networks that operate inside a cell. However to simple the understanding of biological networks and to identify the architecture of the network, networks have been classified based on their molecular components. Design principles associated with each of the networks has also been highlighted to obtain an elaborate understanding on how networks function and can be analysed.

3.2.1 Metabolic Networks

Metabolic reactions are the major source of energy production inside the cell. An enormous amount of products are generated that take part in diverse cellular mechanisms [15]. Metabolic networks therefore comprise of a rich number of enzymes, enzyme-substrate complexes, regulatory proteins and small molecules and their interactions [18]. Metabolic networks therefore define the interactions between the metabolites and the end products. Reactions can be reversible or irreversible, unidirectional or bidirectional, might involve single or multiple species. Based on the kind of reactions, networks can be linear, nonlinear, scalar and scale free networks. Such multifaceted networks contain a variety of graph properties that are comparatively difficult to observe considering the dynamic nature associated with each of the components in the system. Though topology analysis in the steady state gives meaningful insights into the graph properties, however at times the stochasticity of the components needs to be accounted for and hence a probabilistic approach becomes essential. Understanding of such metabolic networks is strongly recommended to understand the energy production in living cells. Several software's are available to visualize metabolic networks such as JDesigner [32], Cell Designer [12], Omix [8], etc. and several software are also available to carry out simulation like COPASI [17]. To understand the dynamics of metabolic pathways and to construct synthetic systems, real time observation is very essential and should be performed with high precision. So far, a number of studies focussing dynamic behaviour of metabolic networks for a variety of organisms have been conducted. Data produced by these studies have paved the way for systems biologists to ascertain underlying design principles moulding the framework of such metabolic pathways. For example, tuberculosis has been a matter of interest for medical science to produce effective drugs against its pathogen (i.e., *Mycobacterium tuberculosis*) and the resistance of disease in humans. Complete genome sequencing of *M. tuberculosis* to understand its biology and detailed mechanism of pathogenesis has provided subsequent clues about the regulatory mechanisms working behind cellular processes such as metabolism, regulation and signal transduction [7]. *M. tuberculosis* contains one of the most enriched metabolic systems in comparison the other pathogens. It can metabolize a number of molecules [13] like lipids and polyketides. It contains an array of enzymes usually found in mammals and other pathogens [7]. Lipid metabolic network found in the tubercle bacillus (Fig. 3.1) is an example of scale-free metabolic network. Metabolic system of *M. tuberculosis* exhibits a range of proteins that work in lipid degradation processes making it capable of intruding mammalian cells. The thorough understanding of the metabolic network and extraction of design principle applied to the network may pave the way for systems biologist in the development of effective drugs for drug resistance tuberculosis.

3.2.2 Transcription Networks or Gene Regulatory Networks

Francis Cricks' saying "DNA makes RNA, RNA makes protein and protein makes us" seems quite understandable in a layman's view. However, when we go into the actual detail of the phenomena of central dogma, we actually come across highly intricate web of non-linear molecular processes and it takes an observational approach to understand the spatiotemporal behaviour of each of the molecules involved.

Transcription is the main course of this abstract orchestra that leads to the formation of most variant and essential machinery of regulatory system of the living cells i.e., proteins. Protein synthesis is regulated at the transcription level by gene regulatory mechanisms. The transcription is controlled by the transcriptional factors (TF's). Transcription factors play an essential role in moderating the production of the proteins that maintain the proper functioning of the cell. Genes and TF's interact with each other to enhance the production of a desired gene product. It is these interactions that are represented in the transcription network [1]. Depending on the requirements, TFs affect the transcription rate of genes per unit time. They thereby act as both repressors and activators of transcription. Bacteria like *E. coli* have highly complex transcription factor networks which are composed of a variety of network motifs and interactions (Fig. 3.2) making it a thousand time difficult to observe the dynamic behaviour of the network [14, 31].

Gene regulatory networks are similar to the transcription networks but they are made of just genes [41]. A gene regulatory network comprises interaction of a gene with other gene leading to the activation or suppression of the activity. Gene regu-

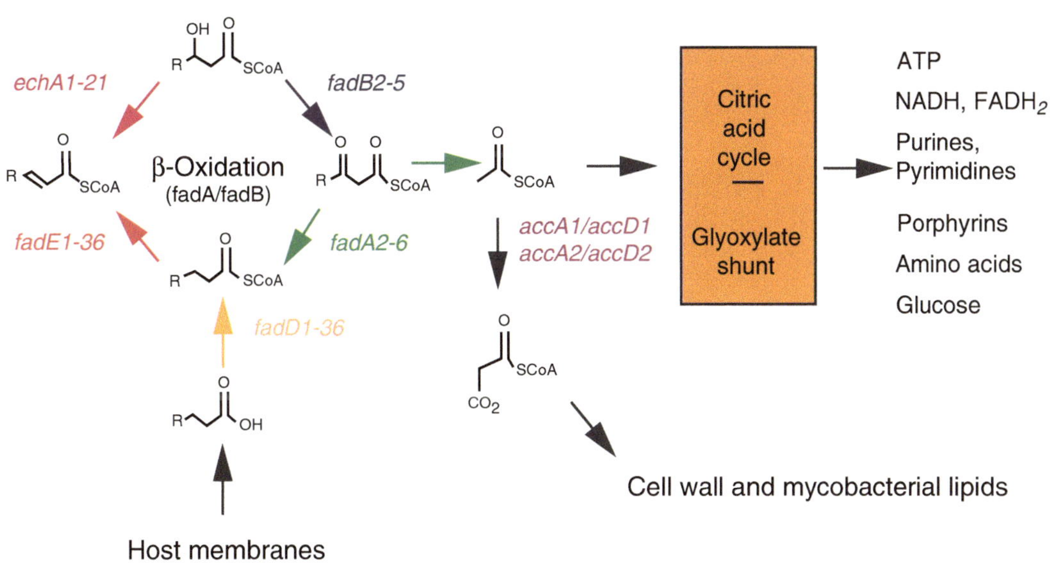

Fig. 3.1 Metabolic pathway of lipid metabolism in *Mycobacterium tuberculosis* showing features of scale free network [7]

lation is also carried out due to extracellular stimuli that the cell receives in its environment in the form of any stimulating factor. Such networks depict the expression level of a gene. Several software are available to visualize gene regulatory networks such as Cytoscape [30], Biotapestry [22], etc.

3.2.3 Signal Transduction Networks (STNs)

Signalling networks depict the underlying structure of cell signalling and how perturbations affect the signal transduction pathways. Understanding the network architecture and dynamic behaviour of the STNs is highly recommended in order to understand cellular systems. To develop more efficient and effective synthetic networks, in depth understanding of signalling networks is a must. Signalling transduction processes are important in the context of cellular sustainability and their response to environmental changes.

STN comprise of a set of specific proteins that work as messengers of external stimuli received by the cell from the environment. Information received by the signalling proteins is then processed and transferred to the internal machinery of the cell. STNs also interact with other networks such as the transcription network, gene regulatory networks, etc. to form even more complex intracellular networks. Several examples of signal transduction networks that are an elaborate depiction of the typical mechanism of signal transduction exist (Fig. 3.3). Several signalling pathways can be modelled into STN for e.g. TNF associated pathway [26], NF-kB pathway [16]. To model STN, several databases are available such as BioCarta [25], NCI database, TRANSPATH [21], etc.

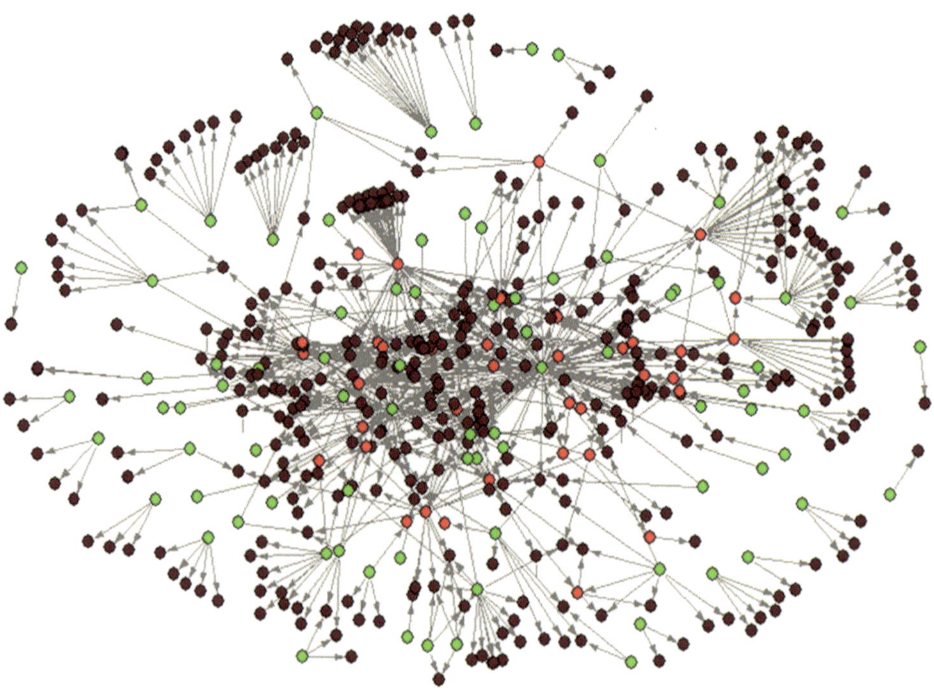

Fig. 3.2 Representation of transcription regulatory network of *E. coli* [13]

3.2.4 Protein-Protein Interaction Networks

Proteins are the most essential part of cellular machinery, which takes part in almost every molecular process inside the cell. Proteins interact with a wide variety of molecular species such as DNA, RNA and other proteins. From our understanding and with the development of molecular biology, it has been relatively easy to derive insights into various protein-protein interactions giving an idea about functioning of various proteins. Proteins also affect the activity of several proteins thereby modulating their functioning. There are several databases that provide information regarding the protein-protein interactions such as DIP [44], BIND [3] and STRING [11]. Computational biologists have developed working strategies to predict functions of uncharacterized proteins using these databases [20, 23, 24, 29]. A protein-protein interaction network is a multi-dimensional graph extending into the direction of interaction of proteins with other proteins. For example, the protein-protein interaction network of all the proteins from *Treponema pallidum* (Fig. 3.4) gives an idea of the complexity of the network that arises from the multi-way interaction of proteins with other

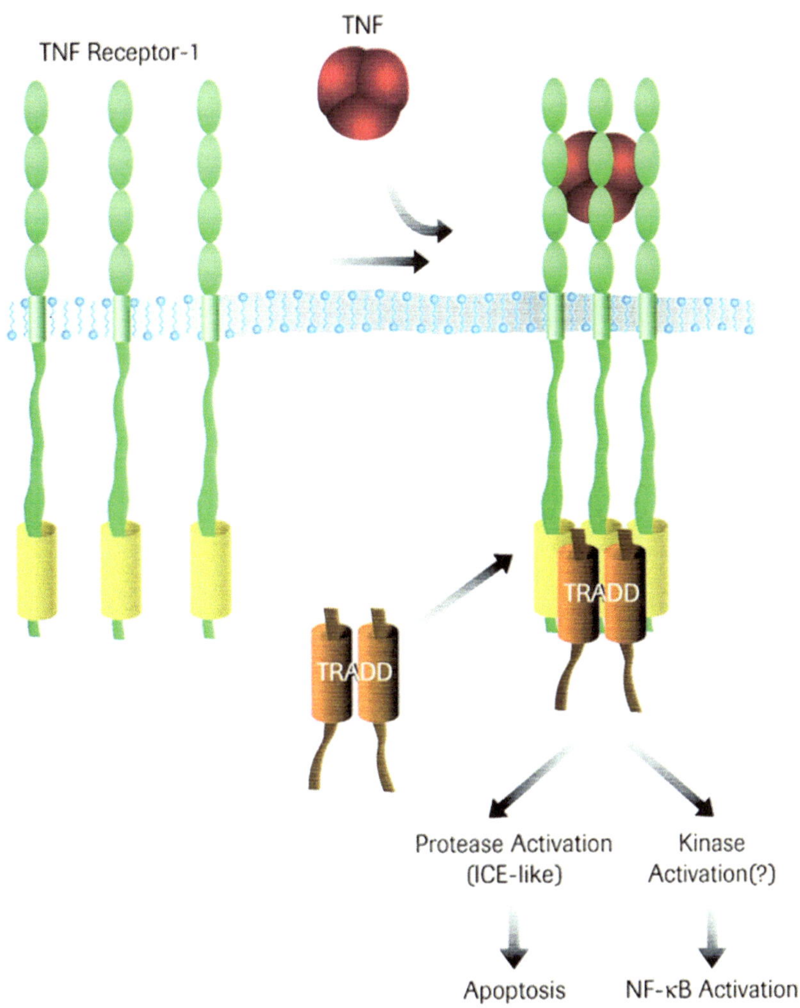

Fig. 3.3 Model showing the activation of two distinct TNFR1 signal transduction pathways by tumor necrosis factor (TNF) [31]

relative proteins. Titz et al. [35] during the study of *T. Pallidum* (Nichols strain) interactome identified 3649 interactions between 726 proteins from the proteome of 1039 proteins. Organism-based network mapping of protein-protein interaction networks may unfold the basic design principles that regulate the phenomena of interactions and their possible effects on other proteins resulting in increase or decrease in the activity.

3.2.5 Protein Domain Networks

Proteins domain networks are defined as the interaction between protein domains arranged in a specific topology to give rise to a certain function [2, 43]. The specific arrangement of protein domains defines their functional specificity. Interconnected domains lose their specific function when their specificity of interaction is lost. There are two kinds of domain-domain interactions i.e. intra-chain domain interactions (interaction between the domain of the same protein) and inter-chain domain interactions (interaction between domains of different proteins). Advances in experimental data depicting clues for such interactions have added a substantial amount in the understanding of the topology and dynamic of such networks. There are established databases which are a repository of such interactions such as DOMINE [45]. Protein domain networks like other complex biological networks show scale-free behaviour such as the domain network of *Saccharomyces cerevisiae* (Fig. 3.5) [42].

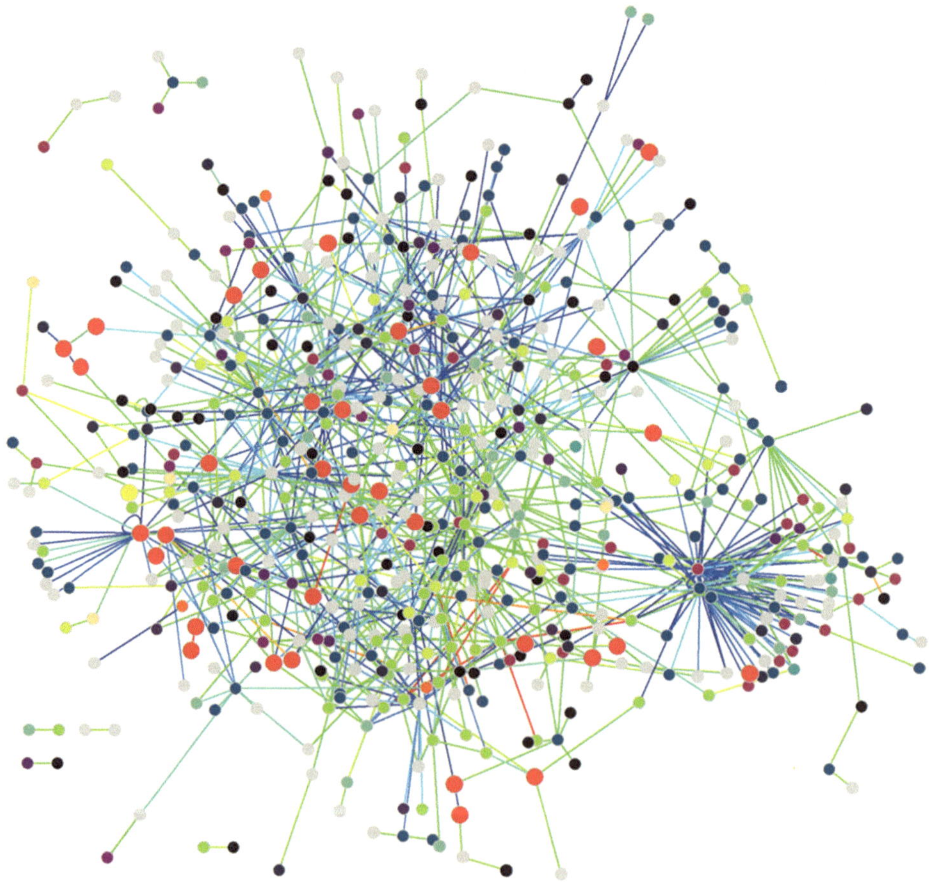

Fig. 3.4 Representation of scale free protein-protein interaction network of the proteins from *Treponema pallidum* (Nichols strain) [35]

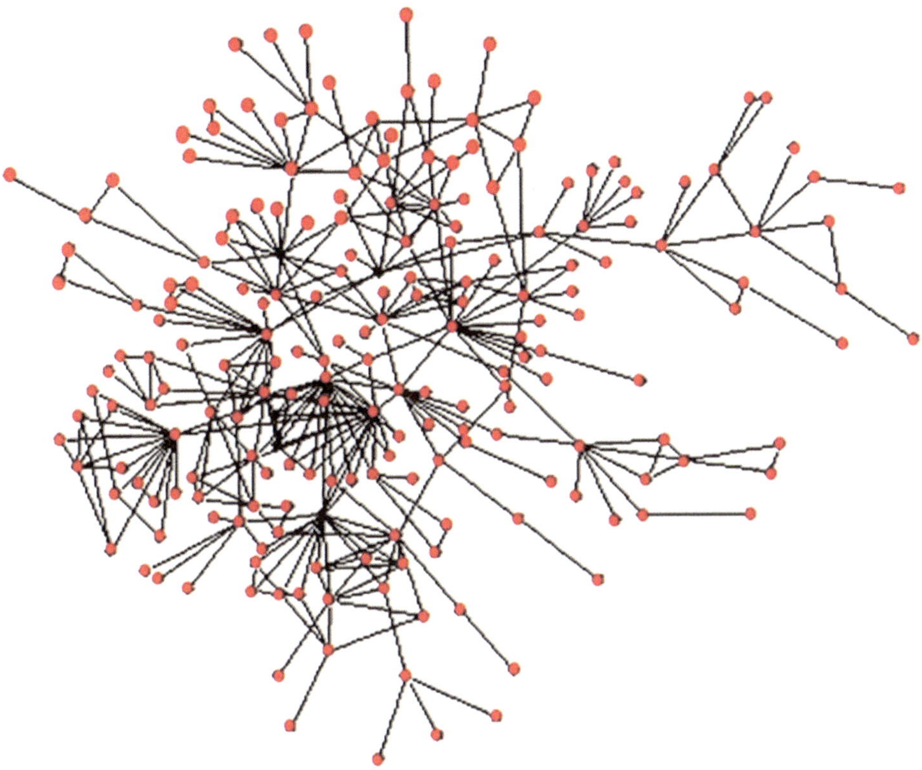

Fig. 3.5 Representation of a major component of the domain network of *Saccharomyces cerevisiae* including 204 vertices and 347 edges [42]

3.2.6 Phylogenetic Trees

Phylogenetic trees qualify for various reasons as biological networks. Phylogenetic trees provide a way to represent biological entities and their interaction in graphical form. From organism level to the molecular level, phylogenetic trees depict an organization of species as hierarchical networks. Hierarchical organization of orthologs and paralogous genes is an explicit example of phylogenetic network. Phylogenetic networks are essential to understand the evolutionary relatedness of organisms and their molecular species. In recent years, genome-based phylogenetic analysis has been in trend to construct phylogenetic observations. These genome-based analysis can be utilized to understand how evolutionary interaction can affect the activity of the molecular species [28]. Phylogenetic networks thereby relate evolutionary pressure that molecular species are subjected to with their functional interactions.

3.3 Analysis Strategies Applied to Biological Systems

In the previous section of the chapter, we have learned about different biological networks, their architecture and underlying design principles behind these networks. In this section, we will try to discuss some analysis strategies developed so far by the system biologists to analyse networks. System, in context of cell as we discussed earlier, is a collection of components (i.e., genes, proteins, transcription factors, etc.) and their relative interactions. By default, every biological system in order to survive against the ongoing perturbations in the environment contains a series of self-regulatory set right systems that help the system to attain robustness. Achieving robustness is the principle objective of any biological system [19]. To understand how at all systems are fabricated, how dynamic a system is and how a system controls itself in order to maintain biological stability are the kind of questions systems biologists have been trying to understanding by analysing biological networks. Following are the widely used analysis strategies to fulfil aforementioned purposes:

3.3.1 Constraint Based Analysis

It is a mathematical approach to study biochemical networks that is capable of dealing with complex networks such as genome-scale metabolic network reconstructions. Flux balance analysis (FBA) analyses the fluxes that operate in the system such that a desired objective is attained for e.g. achieving maximum biomass production. Analysis is based on the stoichiometry of the metabolic reactions wherein the flow of the metabolites of each reaction is represented in the form of mathematical equations [36]. For e.g. Edwards et al. [9] in their study utilized FBA to predict the metabolic capabilities of *E. coli.*

Understanding metabolic networks may lead to an estimate about the capability of metabolite production of a particular organism. While analysing the network, there is no need of any kinetic parameters and models analysed using FBA can give insight into the growth rate of an organism and the rate of production of a specific metabolite that plays a key role in the regulatory mechanism of the organism [27]. In future, more effective models based on FBA may be constructed to acquire control over the metabolic pathways for more complex systems such as humans and other mammals. Another method used for analysing metabolic networks is Metabolic Control Analysis which provides mathematical approach for the understanding of dynamical behaviour of metabolic system [10]. It is useful for understanding the relationship between the steady state properties of biological network and of each of its components. It is kind of sensitivity analysis of a dynamical system. The stoichiometric structure of the network gives an idea of its nature and the control and regulatory mechanisms existing within the network. With the development of even recent techniques such as elementary mode analysis and MOMA, it is expected that future development would relate to the integration of various mathematical analysis methods which would facilitate the generation of more effective and flexible models that can then be used for understanding of several intricate systems.

3.3.2 Bifurcation Analysis

Biological systems can be complex in nature wherein the behaviour of the system can be based on a few of the components or parameters. Bifurcation analysis is a mathematical study of changes in the structure of a particular network with time. System is defined in the form of differential equations wherein it is assumed that bifurcation occurs when a small change is made in some of parameters (also called as the bifurcation parameters). Bifurcations in continuous systems are described in the form of ODE's or PDE's while those in discrete systems are described in the form of maps. Bifurcations can be local or global. In past, several attempts have been made to apply bifurcation analysis for complex biological systems. Borisuk and Tyson [5] applied bifurcation analysis for modelling the mitotic control by M-phase promoting factor (MPF). They introduced several parametric changes to check the feasibility of the model. Bifurcation analysis has remained the primary choice of system biologist while addressing the dynamical behaviour of complex nonlinear systems. Several attempts have been made so far to exploit this strategy effectively. In future, there is scope for successful application of bifurcation analysis to more complex systems.

3.3.3 System Control Analysis

Apart from the extrinsic mathematical analysis strategies applied to biological systems, we find that there exists an array of analysis and control mechanisms such as regulatory mechanism, repair proteins, immune response proteins, and heat shock proteins, etc., in biological systems which work all along to provide stability to the system. In context of system control, two types of control mechanisms ubiquitously found in biological systems are feed forward and feedback control systems. There are several examples of both types of control mechanism distributed in a wide range of biological systems such as feedback control in bacterial chemotaxis and heat shock response which contains both feed forward and feedback control loops. There are a few distinctive examples where both these controls methods are found mutually for example, heat shock response regulation in *E. coli* [6]. The regulation is carried out because of the formation of σ^{32} in response to feedback and feed-forward control mechanism [34]. The understanding of control mechanisms found in other organisms may pave the way for the development of effective control machinery for synthetic biological systems.

3.4 Conclusion

So far, in this chapter we have given a brief overview of historical perspective and systems biology, its approach towards developing system level understanding of biological systems. We have also discussed specifically about how systems are organized into different biological networks such as metabolic networks, transcription networks or gene regulatory networks, signal transduction networks, protein-protein interaction networks, protein domain networksand phylogenetic networks. We have also discussed underlying design principles with the help of elaborative illustrations adapted from various established studies carried out in recent years. This brings us to a conclusion that most of the inbuilt characteristic features of biological networks are governed by simple laws of physics. In the last section of the chapter, we have given an overview of various analytical strategies applied to these biological systems that are both intrinsic as well as extrinsic in nature. Nature has provided biological systems with inbuilt regulatory and repair mechanism meant to control the perturbations in ongoing processes in response to external stimuli such as changes in environmental factors (i.e., temperature, pressure, changes in pH, etc.) or to internal disturbances such as DNA damage, protein misfolding, etc. We have described some other analysis strategies that are applied externally in the form of mathematical models to understand the dynamic behaviour of biological systems.

The principal objective of systems biology i.e. developing an understanding of dynamic behaviour of biological systems is being realized with the help of different fields of science such as electrical engineering, computer science, genetic engineering, genomics, proteomics and transcriptomics. Recent advances in systems biology research have unfolded complex mysteries of dynamics of biological systems with the integration of effective computational methods, simulation techniques and other analysis methods. Data provided by these observations will be helpful for future developments in analysing more complex systems and extraction of design principles to develop efficient systems which will help the process of drug discovery.

References

1. Alon U (2007) An introduction to systems biology: design principles of biological circuits, vol 10, Chapman & Hall/CRC mathematical and computational biology series. Chapman & Hall/CRC, Boca Raton
2. Apic G, Gough J, Teichmann SA (2001) Domain combinations in archaeal, eubacterial and eukaryotic proteomes. J Mol Biol 310(2):311–325. doi:10.1006/jmbi.2001.4776
3. Bader GD, Betel D, Hogue CW (2003) BIND: the Biomolecular Interaction Network Database. Nucleic Acids Res 31(1):248–250
4. Biggs N, Lloyd EK, Wilson RJ (1986) Graph theory, 1736–1936. Clarendon, Oxford/New York
5. Borisuk MT, Tyson JJ (1998) Bifurcation analysis of a model of mitotic control in frog eggs. J Theor Biol 195(1):69–85. doi:10.1006/jtbi.1998.0781
6. Bukau B (1993) Regulation of the Escherichia coli heat-shock response. Mol Microbiol 9(4):671–680
7. Cole ST, Brosch R, Parkhill J, Garnier T, Churcher C, Harris D, Barrell BG (1998) Deciphering the biology of Mycobacterium tuberculosis from the complete

genome sequence. Nature 393(6685):537–544. doi:10.1038/31159

8. Droste P, Miebach S, Niedenführ S, Wiechert W, Nöh K (2011) Visualizing multi-omics data in metabolic networks with the software Omix—a case study. Biosystems 105(2):154–161
9. Edwards JS, Palsson BO (2000) Metabolic flux balance analysis and the in silico analysis of Escherichia coli K-12 gene deletions. BMC Bioinformatics 1:1. doi:10.1186/1471-2105-1-1
10. Fell DA (1992) Metabolic control analysis: a survey of its theoretical and experimental development. Biochem J 286(Pt 2):313–330
11. Franceschini A, Szklarczyk D, Frankild S, Kuhn M, Simonovic M, Roth A, Lin J, Minguez P, Bork P, von Mering C, Jensen LJ (2013) STRING v9.1: protein-protein interaction networks, with increased coverage and integration. Nucleic Acids Res 41(Database issue):D808–D815. doi:10.1093/nar/gks1094
12. Funahashi A et al (2008) CellDesigner 3.5: a versatile modeling tool for biochemical networks. Proc IEEE 96(8):1254–1265
13. Grosset J (2003) Mycobacterium tuberculosis in the extracellular compartment: an underestimated adversary. Antimicrob Agents Chemother 47(3):833–836
14. Guzman-Vargas L, Santillan M (2008) Comparative analysis of the transcription-factor gene regulatory networks of E. coli and S. cerevisiae. BMC Syst Biol 2:13. doi:10.1186/1752-0509-2-13
15. Hatzimanikatis V, Li C, Ionita JA, Broadbelt LJ (2004) Metabolic networks: enzyme function and metabolite structure. Curr Opin Struct Biol 14(3):300–306. doi:10.1016/j.sbi.2004.04.004
16. Hsu H, Shu HB, Pan MG, Goeddel DV (1996) TRADD-TRAF2 and TRADD-FADD interactions define two distinct TNF receptor 1 signal transduction pathways. Cell 84(2):299–308
17. Hoops S, Sahle S, Gauges R, Lee C, Pahle J, Simus N, Singhal M, Xu L, Mendes P, Kummer U (2006) COPASI—a complex pathway simulator. Bioinformatics 22(24):3067–3074
18. Jeong H, Tombor B, Albert R, Oltvai ZN, Barabasi AL (2000) The large-scale organization of metabolic networks. Nature 407(6804):651–654. doi:10.1038/35036627
19. Kitano H (2002) Systems biology: a brief overview. Science 295(5560):1662–1664. doi:10.1126/science.1069492
20. Kumar K, Prakash A, Tasleem M, Islam A, Ahmad F, Hassan MI (2014) Functional annotation of putative hypothetical proteins from Candida dubliniensis. Gene 543(1):93–100. doi:10.1016/j.gene.2014.03.060
21. Krull M, Pistor S, Voss N, Kel A, Reuter I, Kronenberg D, Michael H, Schwarzer K, Potapov A, Choi C, Kel-Margoulis O (2006) TRANSPATH®: an information resource for storing and visualizing signaling pathways and their pathological aberrations. Nucleic Acids Res 34(suppl 1):D546–D551
22. Longabaugh WJ (2012) BioTapestry: a tool to visualize the dynamic properties of gene regulatory networks. Methods Mol Biol 786:359–394
23. Naqvi AA, Shahbaaz M, Ahmad F, Hassan MI (2015) Identification of functional candidates amongst hypothetical proteins of Treponema pallidum ssp. pallidum. PLoS ONE 10(4), e0124177. doi:10.1371/journal.pone.0124177
24. Naqvi AAT, Ahmad F, Hassan MI (2015) Identification of functional candidates amongst hypothetical proteins of Mycobacterium leprae BR4923, a causative agent of leprosy. Genome. doi:10.1139/gen-2014-0178
25. Nishimura D (2001) BioCarta. Biotech Softw Internet Rep Comput Softw J Scient 2(3):117–120
26. Old LJ (1988) Tumor necrosis factor. Sci Am 258(5):59–60, 69–75
27. Orth JD, Thiele I, Palsson BO (2010) What is flux balance analysis? Nat Biotechnol 28(3):245–248. doi:http://www.nature.com/nbt/journal/v28/n3/abs/nbt.1614.html#supplementary-information
28. Pellegrini M, Marcotte EM, Thompson MJ, Eisenberg D, Yeates TO (1999) Assigning protein functions by comparative genome analysis: protein phylogenetic profiles. Proc Natl Acad Sci U S A 96(8):4285–4288
29. Shahbaaz M, Hassan MI, Ahmad F (2013) Functional annotation of conserved hypothetical proteins from Haemophilus influenzae Rd KW20. PLoS ONE 8(12), e84263. doi:10.1371/journal.pone.0084263
30. Shannon P, Markiel A, Ozier O, Baliga NS, Wang JT, Ramage D, Amin N, Schwikowski B, Ideker T (2003) Cytoscape: a software environment for integrated models of biomolecular interaction networks. Genome Res 13(11):2498–2504
31. Shen-Orr SS, Milo R, Mangan S, Alon U (2002) Network motifs in the transcriptional regulation network of Escherichia coli. Nat Genet 31(1):64–68. doi:10.1038/ng881
32. Sauro H (2004) An introduction to biochemical modeling using JDesigner. Keck Graduate Institute, Claremont
33. Spivey A (2004) Systems biology: the big picture. Environ Health Perspect 112(16):A938–A943
34. Straus DB, Walter WA, Gross CA (1987) The heat shock response of E. coli is regulated by changes in the concentration of sigma 32. Nature 329(6137):348–351. doi:10.1038/329348a0
35. Titz B, Rajagopala SV, Goll J, Hauser R, McKevitt MT, Palzkill T, Uetz P (2008) The binary protein interactome of Treponema pallidum-the syphilis spirochete. PLoS ONE 3(5), e2292. doi:10.1371/journal.pone.0002292
36. Varma A, Palsson BO (1994) Metabolic flux balancing: basic concepts, scientific and practical use. Nat Biotechnol 12(10):994–998
37. Von Bertalanffy L (1950) The theory of open systems in physics and biology. Science 111(2872):23–29

38. Watson JD, Crick FH (1953) Molecular structure of nucleic acids; a structure for deoxyribose nucleic acid. Nature 171(4356):737–738
39. Weiner N (1948) Cybernetics or control and communication in the animal and machine. MIT Press, New York
40. Westerhoff HV, Palsson BO (2004) The evolution of molecular biology into systems biology. Nat Biotechnol 22(10):1249–1252. doi:10.1038/nbt1020
41. Winterbach W, Van Mieghem P, Reinders M, Wang H, de Ridder D (2013) Topology of molecular interaction networks. BMC Syst Biol 7:90. doi:10.1186/1752-0509-7-90
42. Wuchty S (2001) Scale-free behavior in protein domain networks. Mol Biol Evol 18(9):1694–1702
43. Wuchty S (2002) Interaction and domain networks of yeast. Proteomics 2(12):1715–1723. doi:10.1002/1615-9861(200212)2:12<1715::AID-PROT1715>3.0.CO;2-O
44. Xenarios I, Salwinski L, Duan XJ, Higney P, Kim SM, Eisenberg D (2002) DIP, the Database of Interacting Proteins: a research tool for studying cellular networks of protein interactions. Nucleic Acids Res 30(1):303–305
45. Yellaboina S, Tasneem A, Zaykin DV, Raghavachari B, Jothi R (2011) DOMINE: a comprehensive collection of known and predicted domain-domain interactions. Nucleic Acids Res 39(Database issue):D730–D735. doi:10.1093/nar/gkq1229

4 Structureomics in Systems-Based Drug Discovery

Lumbini R. Yadav, Pankaj Thapa, Lipi Das, and Ashok K. Varma

Abbreviations

SBDD	Structure-based drug designing
ADME	Absorption distribution, metabolism, and excretion
SEM	Scanning electron microscopy
TEM	Transmission electron microscopy
XRD	X-ray diffractometer
PSI	Protein structure initiative
SGC	Structural Genomics Consortium

4.1 Introduction

Structure biology deals with the study of three dimensional structures of macromolecules like proteins, DNA, and RNA. The target molecule for structural study is protein, a string of amino acids which fold into loops, secondary, tertiary, and quaternary structures. Structural studies of these molecules reveal the 3D atomic level details, effect of mutations on protein folding and function. Furthermore, the use of *in silico* bioinformatics-based approach has helped to determine the 3D structure of proteins from primary sequence [1]. High-resolution structure of protein helps in understanding the protein dynamics, protein folding, and structure-guided functions of proteins. The experimentally determined structures of protein molecule are useful in molecular modelling and computational biology studies. Structure of different molecules like DNA, RNA, proteins, and their complexes with ligand are also reported from different organism [2]. These structures had till date played a very important role in structure-based- drug designing.

Biology, which includes the study of living organisms, has become abundantly rich with data obtained from number of biological studies, experiments, and also due to recent advancements in technology. This outburst of information led to an emergence of a new field called "OMICS". Omics is the study of biological molecules of an organism that perform different functions. Omics aims at comprehensive characterization and quantification of biological molecules that are present in the organism/organisms. Omics is attached to different prefix which describes the field of studies, for example, the study of genome is known as genomics, study of proteome is known as proteomics, and so on. Different field of omics study include lipidomics, transcriptomics, metabolomics, interactomics, stem cell genomics, and structural proteomics.

Author Contributions
Conceived and designed the chapter: AKV Performed the analysis: LYR, PT, LD, Wrote the chapter: LYR, AKV.

L.R. Yadav • P. Thapa • L. Das • A.K. Varma (✉)
Varma Lab, Advanced Centre for Treatment, Research and Education in Cancer, Kharghar, Navi Mumbai, Maharashtra 410 210, India
e-mail: avarma@actrec.gov.in

S. Singh (ed.), *Systems Biology Application in Synthetic Biology*,
DOI 10.1007/978-81-322-2809-7_4

Study of omics is useful to identify different molecules present in the organisms, evolution of organism, orthologous and paralogous genes present in the organism, and novel regulatory processes present in the organism at transcription and translational levels. Study of metabolomics and structureomics will play a significant role in the process of drug discovery.

Metabolomics can be an invaluable tool for clinical studies like drug toxicity, early diagnosis of preclinical conditions, and identification of biomarkers. Structural proteomics is the study of structural aspects of whole cellular components which aims at (1) determining the 3D structures of diverse subset of proteins which can be used to model other structures using computational techniques and (2) mapping the structures of proteins and protein-protein interactions from a large number of model organisms.

Eventually, the goal lies in strengthening the computational methods so that reasonable structures for every sequence can be determined at high resolution. Structural proteomics will help to computationally generate or experimentally determine and view the 3D structure that correlates with protein function. The 3D structure of proteins obtained can provide molecular insights of the proteins that can be used as druggable targets for designing the small molecule inhibitors against various diseases and interfere with resistance development in organisms.

4.2 Drug Discovery

Drug discovery is a process of identifying small molecules which can bind and modulate the function of a target molecule. Proteins are involved in myriad of cellular processes making them effective drug targets. Drug discovery and design requires the identification of potential drug candidates, novel target and characterization followed by biochemical assays to test their therapeutic efficacy. The drug discovery process is often lengthy, difficult, and expensive. The discovery of drug involves a multidisciplinary effort of scientists and clinicians to explore the new approaches for therapeutics. The major steps in the process of drug discovery include: (1) Identification of a disease associated specific molecular target; (2) Identification of hits and leads (small molecule inhibitors, monoclonal antibodies) to intervene with the molecular target for reversal or inhibition of the disease; (3) Understanding the detailed 3D structure of the target with lead compounds that affect the function; (4) Optimization of the lead compounds to increase the efficacy and potency that is further examined in preclinical studies. The different steps in drug discovery can be broadly divided into different subheadings as follows.

4.2.1 Investigation of Drug Target and Lead Molecules

Understanding the biology of a disease gives new insights about the molecules that can be targeted for drug development or diseases. The aim for drug design is to identify a biological target and ligand molecules that can act as a promising inhibitor/promoter, etc. The identified targets and drug leads are further validated, and the lead is optimized to enhance its potential benefits and mechanisms of action.

4.2.1.1 Target Identification and Validation

A target is a biological entity which elicits a biological response that can be measured experimentally on binding to a drug molecule. A few basic criteria are to be considered before selecting the molecule for drug discovery: (1) The target molecule should be indispensable for the survival of the cell; (2) The drug molecule should specifically target to the protein or protein pathways; (3) The protein should have a small-molecule binding site for which a compound can be designed; (4) The target molecule 3D structures should be determined and its best to have co-crystallized structure with inhibitors. G-protein-coupled receptors (GPCRs) are known to be more responsive to small molecule drug whereas antibodies are good at interfering with protein-protein interactions [3]. Target can be identified through examining the correlation of protein levels with disease progression, genetic polymorphism and the risk of disease, and isolation of monoclonal antibodies that bind tumor cells [4]. The identified target is then validated using *in vitro tools*, animal models, and study of desired target in patients. Recently, the field of chemical genomics has emerged that studies the genomic response in individuals when challenged with chemical compound. The aim is to provide a chemical tool against every protein transcribed and translated [5].

4.2.1.2 Hit to Lead Identification

In general, the molecule which is to be considered as a drug molecule should obey the Lipinski's rule of 5 [6]. Lipinski's rule of five considers orally active compounds that have achieved phase II clinical status and defines four simple physicochemical parameter ranges (MWT $\leq$ 500, log P $\leq$ 5, H-bond donors $\leq$ 5, H-bond acceptors $\leq$ 10) associated with a drug. Previously, in vitro screening was performed to identify lead compounds and focus was to find drug-like compounds more than lead-like compounds [7]. The optimization of leads within the Lipinski's rule may be difficult [8]. This led to a pioneering work called as "SAR by NMR" (Structure Activity Relationship by Nuclear Magnetic Resonance) method that screens smaller and simpler molecules for the discovery of lead. The process of generating lead compounds is through a fragment-based screening and diversity oriented screening [9, 10]. Once the hit molecule is identified and optimized for the strong affinity interaction, its co crystal structure with the ligand can be obtained. The information from these co-crystals will help in mapping the binding site of the target and also help in further optimization of the compounds identified. A variety of ways exist to identify hit molecules for further lead development and optimization.

Structure-Guided Drug Discovery

Structure-guided drug design method utilizes the information from the 3D structures of the target molecules, the ligand, or the target-ligand complex for drug discovery. The ligand target interface provides in depth information about molecular orientation between the interacting groups, the number and strength of hydrogen bonds, hydrophobic interactions, the presence of water molecules, or any ionic atom at the active site. The definition of topographies at the interaction surface of the ligand and target helps to optimize the potency and selectivity [11]. 3D structural information till date has played a major role in drug discovery for several classes of drug targets. As membrane proteins are difficult to crystallize, novel approaches for the 3D structure determination of integral membrane proteins by solution NMR are in progress [12]. Lopinavir, a potent second-generation HIV-1 protease inhibitor, was synthesized using structure-based design of HIV-1 protease. Lopinavir is effective against mutants resistant to Ritonavir. The success of Lopinavir is based on the crystal structure of complex HIV-1 protease and Lopinavir [13].

Looking at the importance of structure-guided drug design, it is important to keep in mind the limitations of this method. Artifacts introduced during crystallization, structure refinement, and structure solution can have substantial influence when such structures are used for drug design, docking, and virtual screening [14, 15]. Crystallization conditions of the protein, change in conformation of protein in different buffer conditions, distortion in crystals due to soaking in ligand, interference of ligand binding due to crystal packing, and crystal packing that drives the ligand binding are all the problems associated with the SBDD method.

Alteration made in the protein to increase the probability of crystallization and low resolution structures can also affect the SBDD [32, 33]. Low resolution structures incorporate uncertainty in the atomic position (for 3 Å structure an error of 0.5 Å in the position of individual atoms) [14, 34]. This uncertainty is critical in an inhibitor design program, since both hydrogen bonding and hydrophobic interactions are very sensitive to distance and direction and also important for drug designing [35, 36]. Table 4.1 shows a few examples of the drugs designed using SBDD.

Table 4.1 Drugs discovered by structure guided drug design. List of few examples of drugs discovered from SBDD, their molecular targets, and the disease for which it is used

Drug	Protein target	Disease
Zelboraf	Serine threonine protein kinase BRAF	Melanoma [16]
Gefitinib	EGFR inhibitor	Non-small cell lung cancer [17]
Agenerase/Viracept	HIV protease	AIDS [18, 19]
Gleevec	BCR-ABL	Chronic myelogenous leukemia [20]
Tarceva	ATP-binding site of EGFR	Non-small cell lung cancer [21]
4MCHA and AdoDATO	Spermidine synthase	Malaria [22]
Relenza	Neuraminidase	Influenza [23]
Canertinib	Epidermal growth factor receptor kinase	Cancer [24]
Methotrexate	Dihydrofolate reductase	Megaloblastic Anemia [25–27]
AG-7088	Rhinovirus 3C protease	Common cold [28]
Zonisamide	Human carbonic anhydrase II	Seizures [29]
Prinomastat	Matrix metalloproteinase	Non- small cell lung cancer [30]
Lidorestat	Aldose reductase	Chronic diabetic complications [31]

Computer-Aided Drug Design

Computer-aided drug design methods screen virtual compound libraries against protein target with a known 3D structure. The structural details at protein ligand interface enables to engineer the physico-chemical characteristics of the ligands. This helps in designing focused compound libraries. The energy of interaction between the ligand and target interface helps in sorting of identified hit based on their binding affinity. Modifications on the structure of hits obtained may improve the binding affinity and other properties of lead compounds. This process of hit expansion, lead generation, and optimization may result in a potent lead molecule. The advantage of *in silico* screening method makes it possible to screen large number of compounds in less time and cost. In the computer-guided method of drug discovery, certain issues like structural water interactions, protein flexibility, small-molecule initial geometry, and the scoring and ranking of docked molecules need to be addressed to increase the reliability of the output. MASC (multiple active site correction), a novel scoring method, addresses some of the limitations with current methods [37]. Molecules identified by *in silico* methods are further evaluated and validated for the binding of the lead molecule using biophysical screening methods, like thermal-shift assay, nuclear magnetic resonance (NMR), and X-ray crystallography.

Fragment-Based Drug Design/Discovery

Fragment-based drug discovery involves screening of low molecular weight fragment libraries (<250 Da) directed against a target of interest. The fragments selected for screening are filtered for characteristics that include lipophilicity indices, higher ligand efficiency, and exploration of chemical diversity in space, exclusion of reactive or metabolically active groups. This screen therefore offers a greater likelihood of finding hits useful for lead discovery [38]. The strategy used in fragment-based drug discovery to modify the fragment molecules are privileged for fragment-based reconstruction approach [39–41], fragment hybridization based on crystallographic overlays to create a new hybrid compounds with enhanced affinity and efficacy [42, 43], fragment growth exploiting dynamic combinatorial chemistry [44, 45], and high-speed fragment assembly via diversity-oriented synthesis followed by *in situ* screening bids a way for more efficient and rapid discovery of novel drugs [46, 47]. Biophysical methods and *in silico* techniques have proved useful in fragment-based drug discovery to identify molecules that bind with high affinity to target and add only a small entropic penalty. The sensitive biophysical methods used to screen and validate fragment binding include nuclear magnetic resonance, isothermal titration calorimetry, surface plasmon resonance, and differential scanning fluorimetry. The experiences of last few decades of hit to lead development and further study of drug candidate in clinical trials indicated that the combination of fragment-based drug discovery and structure-based drug design is more superior to "traditional" methods of drug discovery [48].

Scaffold-Based Drug Discovery

Scaffold-based drug discovery methods screen libraries of around 20,000 compounds with molecular weight in the range 125–350 Daltons. Biochemical methods and co-crystallography are used as the primary screening approach. It involves three steps – scaffold identification, scaffold validation, and chemical optimization. In this method, bioactive compounds co-crystallized with the target are used for further optimization of the lead molecule to increase the bioactivity and affinity.

De novo Structure Determination of Ligand

In this method, structure of ligand is built on the basis of binding affinity by introducing small functional groups. These structures are then docked into the binding site of target, followed by energy minimization and then manually modified by linking the chemical fragments to make the lead compounds [49–51]. Alternatively, core structures can also be derivatized with different functional groups considering the physicochemical characteristics of the binding site [52]. *De novo* ligand synthesis also utilizes "scaffold hopping" approach and information from known ligands through hybridization and/or linking of the input structures [53].

4.2.2 Preclinical Research

Preclinical development generally involves understanding the effect of drug distribution, metabolism, and toxicity. The lead molecules are tested for their pharmacokinetic, pharmacodynamics, ADME (absorption, distribution, metabolism, and excretion), and toxicity. Typically, both *in vitro* and *in vivo* tests are performed. The lead molecule that shows promise as a therapeutic agent is further characterized for its size, shape, toxicity, and bioactivity. Drug formulation, delivery, and packaging are refined continuously to determine the drug's stability for all the parameters involved with storage and shipment, such as heat, light, and time.

4.2.3 Clinical Research

A clinical trial is a research study carried out to understand the efficacy, safety, and effectivity during the treatment of medical technology. These interventions may be from new available medicine/drug, medical device, new therapies, vaccines, or even new ways of using already established treatments. In clinical trials, the effects of drugs under investigations are studied and also are compared with patients treated with already existing drugs in the market. There are different kinds of clinical trial that exists depending on the overall aim of the researchers and clinicians (Table 4.2). Clinical trials are of different

Table 4.2 Types of clinical trials

Sr. No	Types of trials	Goals
1	Interventional trials	Participants take an experimental new drug or undergo surgery
2	Prevention trials	Explore better ways to prevent disease include lifestyle changes or use medicines, vaccines, vitamins, and minerals deficiency of which could predispose the individual
3	Observational trials	Epidemiological survey. Family histories or biological fluids are tested for the survey
4	Screening trials	To determine the best way to detect certain diseases or health conditions
5	Quality of life trials (or supportive care trials)	To search for ways to improve the quality of life for individuals with a chronic illness

kinds and are assigned four main clinical development phases http://www.fda.gov/Drugs/ResourcesForYou/Consumers/

4.2.3.1 Phase I Trials

Phase I trials determines the safety and tolerability of drugs in healthy volunteers. Volunteers of about 20–50 are examined for duration of few minutes up to 2 weeks. Various pharmacokinetic parameters like absorption, distribution, metabolic breakdown, and excretion at different dosages are monitored. The interactions of drug with the food and other medicines taken simultaneously are monitored.

4.2.3.2 Phase II Trials

Patients with the specific illnesses are investigated with the drug under study. Clinical effect and doses are optimized on few hundred patients and treatment is normally monitored for not more than 3 months.

4.2.3.3 Phase III Trials

Phase III trials monitor the safety and efficacy of drugs on large number of patient populations over an extended period of time. This phase includes several thousand patients and the treatment duration and monitoring can be up to a year or longer. The data obtained from these trials are provided to the regulatory authorities of pharmaceuticals to determine whether the drug can be marketed as medicine.

4.2.3.4 Phase IV Trials

The effect of drug is investigated for further validation. In this phase, the focus is to compare or use in combination with other established drugs to generate more data on safety under broader use. It is important step to strengthen the understanding of the drug and to give guidance for the safe and appropriate use under various clinical conditions. Phase IV trials are by definition always performed on the approved drugs, the number of patients can be both small and also extremely large (10–30,000 patients) [54–56]. Figure 4.1 shows overview of steps involved in the process of drug discovery.

4.3 Structureomics

The determination of 3D structure of a protein, at atomic level on a genome-wide scale, to understand the association of sequence with structure and function is known as structural proteomics. Although in literature, the terms "structure proteomics" and "structural genomics" is used interchangeably, "structural proteomics" may be more accurate [57]. Here "Structureomics" refers to the word 'structural proteomics'. Comprehensive survey of the US FDA's *Orange Book* and Centre for Biologics Evaluation and Research (CBER) website, which report for small molecular and biological drugs, have shown that only 1357 unique drugs were present. Of these, 166 were biological drugs and 1204 were small-molecule drugs.. All these drugs are known to act through 324 distinct molecular targets, out of these 266 are human genome derived protein. The current available drugs targets approximately 130 druggable domains most of which belong to four key gene families: class I GPCRs, nuclear receptors, ligand-gated ion channels, and voltage-gated ion channels [58]. Recent advancements like high-throughput crystallization methods, multiple-wavelength anomalous dispersion (MAD), synchrotron beam lines and robotics, and automated crystallization methods have provided remarkable breakthrough in high-throughput structural biology [59–62].

4.3.1 Proteins: The Basic Executor of the Cell

Proteins are the highly complex molecules that drive essential bioprocesses in the cell. The diversity of the protein at the amino acid sequence level and certain post-translational modifications add to the difficulty in understanding the protein functions. Proteins change their conformation by interacting with their binding partners and perform different function. The post-translational modifications like phosphorylation, glycosylation, carbonylation, methylation, and ubiquitination play crucial roles in regulating complex

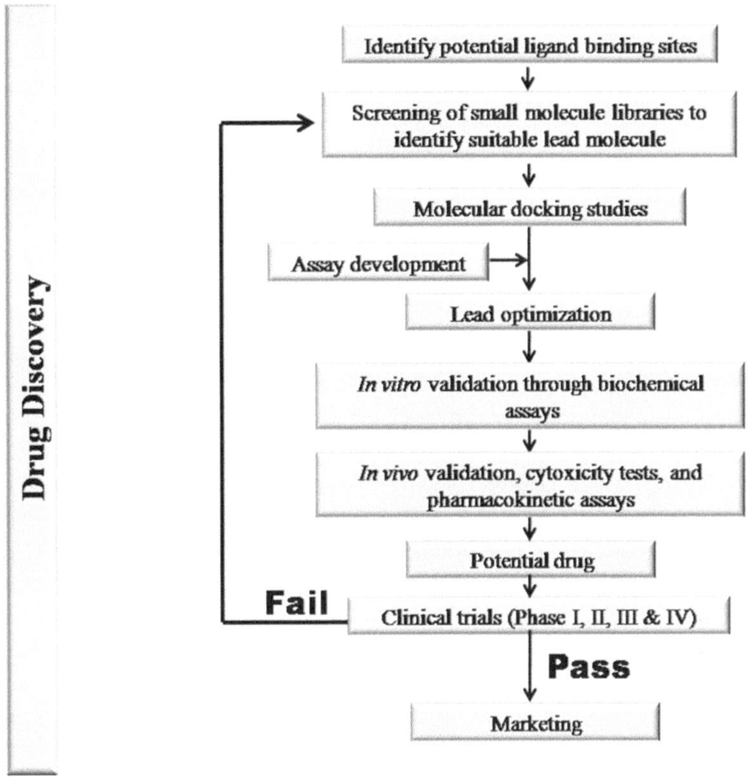

Fig. 4.1 Classical drug discovery pathway from target selection, through lead discovery to lead optimization and finally as a drug candidate. An average drug discovery process requires at least 10 years with billions of rupees invested in the entire process

processes in the cell [63]. They also form simple to large complexes to monitor and accomplish the different task in the cell. Purification and crystallization of membrane protein also poses major challenges. Structure determination of the purified membrane protein will be a feasible goal with the advancement in cryo-electron microscopy.

Proteomics and structureomics study are important to unravel the complexities that we encounter in understanding the functions of biomolecules. Recent studies have revealed the multiple roles for the RNA that has in the various regulatory process of cell.

4.3.2 Methods in Structural Proteomics

The genome of around 100 different organisms including archaea and bacterial species, nematode, fruit fly, rice, and humans have been sequenced, and the growth of sequenced genome in the databases is rising exponentially [64]. However, a large set of proteins translated from the sequences of these genes are not annotated. Researchers have always strived to get maximum information of proteins with regard to their structure and functions using computational approaches. This has been popularized greatly due the availability of sequences and protein structures in the public domain. The information about the sequences from these databases can be used to predict the function and structure of an unreported protein having similar sequence to reported proteins.

4.3.2.1 Function Basis From Primary Sequence of a Protein

Sequence Comparison or Homology-Based methods: Sequence homology is similarity between sequences or degree of similarity between sequences. This similarity in sequences of polypeptide of a protein is indicative of the fact that they may have structural, functional, or evolutionary relationships, and such similar sequences are called homologous sequence. The comparison is done by aligning the unknown sequence with a reference database or known sequence. This alignment is done using different programs like Clustal W [65], LALIGN, and BLAST [66]. However, rearrangements are done in order to span the entire length of the query sequence by giving penalties to each gap inserted. These programs use algorithms that assign a score depending on the sequence similarity or similar physicochemical properties and gap penalties. The confidence with which these alignments are done is very critical for other algorithm or software used for predicting the functions from the sequences. This process is error prone and also amplifies the error since a wrongly annotated protein will lead to misguided functions.[67] *Sequence Motif-Based Method*: Protein motifs are stretch of amino acids sequences that may have functional or biological significance. For example, GGXGXD (where X stands for any amino acid) is a motif present in some metalloproteases that binds to calcium ions and stabilizes the protein [68]. The protein molecules can perform their functions through few of ~ 10 amino acid residues that are present in the binding and catalytic sites [69]. This stretch of amino acids in the active sites has a signature pattern which is nothing but the motif that is associated with a particular function. Protein Families Databases (Pfam) [70] PROSITE [71], BLOCKS [72] and PRINTS [73] are the few examples of motif searching database. Apart from the bioinformatics approaches, microarray analysis, yeast two hybrid system, enzyme activity assays, knock down- knock out studies in animal models, and RNA interference are also useful to establish the function of proteins.

4.3.2.2 Structure Prediction From Sequence

A protein attains its native form by a series of conformational changes, where the primary sequence folds to form the secondary structure, which on further folding forms the tertiary and quaternary structure. Protein sequences, as a template, are not only used for predicting the function but are also used for structure prediction. The strategies for structure prediction from sequence include comparative modelling, fold recognition, and *ab initio* modelling methods [74]. Comparative modelling is also known as homology modelling and as the name suggests it compares the query sequences and aligns it with the known structure. Alignment can be local or global, where a short stretch of the sequence or the entire sequence, is aligned and compared. SWISS-MODEL is one such server which uses comparative modelling method to predict the structure [75]. Fold recognition method uses proteins with known folding pattern as a template. *Ab initio* modelling is a tedious and crude method. An ab initio modelling attempts to build the structure from scratch (using only the sequence information) and conducts a conformational search. This method usually generates a number of all possible conformations that could be attained by the protein. Then it assigns energy function to get minimum potential energy structure that is more thermodynamically stable. These stable structures are closest to the native structure of the protein. It is used when comparative and fold recognition methods fail to identify similar protein with known fold. This is because *ab initio* modelling method only relies on the primary sequence of the protein [76]. There are various other software tools that can be used to predict the structure from the sequence (Table 4.3).

4.3.2.3 Structure Information for Functional Annotation

Determining the structure of the protein is just a part of techniques. The next challenge is functional annotation of the protein. There are many proteins which have similar structures but distinct functions and vice versa, which makes it necessary to correctly annotate the function of a protein. Several methods for predicting the function of a protein have been classified on the basis of their spatial structure which imparts specificity. These spatial regions in the proteins are analyzed by overall folding of proteins critical for the function [77]. The first step in functional annotation of a protein on the basis of structure involves finding a fold match which can be performed by different software like DALI (uses algorithm for pair-wise alignment of protein structures) [78], SSM (uses graph theory) [79],

Table 4.3 Software used to predict structure from protein sequence. Most of them use either, the *ab initio* or the comparative modelling approach to predict structures

Sr. No	Software	Method used	Description
1	Raptor X	Comparative modelling	Carries out 3D structure and binding site prediction
2	I-TASSER	*ab initio* modelling and fold recognition	Predicts both function and structure
3	Robetta	Comparative and *ab initio* modelling	Predicts tertiary structure
4	Modeller	Comparative modelling	Predicts structure by minimizing the spatial restraints
5	Phyre & Phyre 2	*ab initio* modelling	Uses multitemplate alignment protocol

and VAST (uses vector alignment of secondary structures) [80]. Lower level of folds (which can be surface clefts or pocket binding regions) also holds important information about the function of the protein. Structural clefts or pocket regions can be compared in databases like pvSOAR (detect similarities in surface clefts and compares pocket across different proteins) [81] and SURFACE (annotates surface patches based on the structure and sequence information derived from interaction studies) [82]. PreDS (uses electrostatic potential information of the surface to detect DNA binding sites) [83] and NPDock (uses docking and refinement steps to obtain best promising solutions) [84] servers are used for predicting the docking sites of proteins.

4.3.2.4 Protein Production

Protein production and purification is an essential prerequisite to study the structureomics. The large amount of protein can be further used in the commercial production of enzymes, nutritionally valuable proteins, and biopharmaceuticals and most importantly for drug design. After selecting the protein to be purified, its cDNA is cloned into an appropriate expression vector, which is then transformed into suitable host cell. The protein thus over-expressed is called a recombinant protein. The most popular system for protein production is the prokaryotic system *Escherichia coli*, which has been genetically engineered to produce different strains which help to overcome the initial problems faced during protein production, like degradation of recombinant proteins by proteolytic enzymes, leaky transcription, and codon bias. Some of the widely used protein expression strains include BL21 (DE3), Rosetta 2 (DE3), and BL21 Star (DE3) pLysS E. Eukaryotic expression systems like insect cell lines and yeasts are also used which are comparatively costly, time consuming, low yielding, and tedious. Mammalian cell lines used for protein production include HeLa, HEK293T, U2OS, A549, NIH 3 T3, L929, HEK 293, MCF-7, and Hep G2 [85]. Recently cell free protein expression systems have been developed which contains transcriptional, translational, and posttranslational modification machinery needed for *in vitro* protein production. Although these cell-free systems are simpler, they cannot be used for large-scale protein production.

Once the recombinant protein is expressed, purification can be achieved by several techniques, depending on the physical and chemical properties of the protein. The solubility of the protein is an important aspect to be considered during different stages of purification. Insoluble proteins sometimes form inclusion bodies which are difficult to purify. Soluble proteins, on the other hand, can be harvested from the cell lysate by centrifugation. The protein of interest is then separated on the basis of their solubility, size, charge, binding affinity, etc. For the ease of purification, these recombinant proteins are tagged with affinity tags (GST, 6xHis, and MBP). Choice of the affinity column depends on the type of tag present in the vector. Highly purified proteins are obtained by additional steps, which generally include gel permeation or ion exchange chromatography. The purified protein thus obtained can then be confirmed for its identity by peptide mass fingerprinting or western blotting.

4.3.3 Techniques for Structure Determination

The different structure determination techniques include X-ray crystallography, nuclear magnetic resonance spectroscopy, and cryo-electron microscopy [86]. Recent developments of new structure determination techniques include neutron diffraction, fiber diffraction cryo-EM tomography, correlative microscopy, X-ray imaging, single molecule techniques and in-cell NMR. Other approach that is used to determine the structure is through understanding of bioinformatics that look for patterns among the diverse sequences that give rise to a particular shape. The detailed high resolution structure of a protein molecule is useful in designing small molecule inhibitor that has the potential as a pharmaceutical compound. The atomic level details of a molecule is also useful to modify drugs with specific changes to increase the drug efficacy. Out of all the structures submitted in the Protein Data Bank (PDB), over 80 % have been solved using X-ray crystallography, 16 % are solution NMR structure, and 2 % by using theoretical modelling [87]. Different techniques with brief information are tabulated (Table 4.4).

4.3.3.1 X-ray Diffraction

X-ray crystallography is a tool used to determine the 3D position of each atom present in the crystal lattice of the protein crystal. It is the only technique that is being used to solve the structure of the molecule at a resolution of better than 1 Å. The major bottleneck in structure determination using X-ray is obtaining an optimum sized protein crystal. The buffer used for protein crystallization mainly consist of a buffering agent, precipitant, and salt. The most widely used precipitants include PEG (of varying molecular weight), ammonium sulfate, and some alcohols which when combined with other additives give various permutations and combination of buffers. For high-throughput crystallization, screening different robotic facilities are also available. The protein molecules in the crystals act as a signal amplifier as they are aligned in a crystal lattice and diffract the X-ray. The diffraction pattern obtained is analyzed for structure factor which is used to build the electron density of atom. The details thus obtained are based on all the complex calculations, probabilities, and assumption, and it needs to be established as the accurate or the closest to the accurate structure by refining the model at several steps. The accuracy of the model obtained after rigorous refinement is measured with regard to the R-value [88].

4.3.3.2 Nuclear Magnetic Resonance Spectroscopy (NMR)

Nuclear magnetic resonance spectroscopy is another technique to elucidate the solution struc-

Table 4.4 Table shows different experimental approaches to elucidate the structure of the protein with varying resolution. X-ray crystallography has contributed 80 % of all the solved structures in Protein Data Bank (PDB)

Sr.No	Techniques	Principle	Sample
1	Macromolecular crystallography	Diffraction of X ray beam	Crystals
2	Nuclear magnetic resonance spectroscopy of proteins (NMR)	Interaction between an applied magnetic field and the nuclei of certain atom inside proteins	Protein solution
3	Cryo-electron microscopy (cryo-EM)	A beam of electrons in an electron microscope, creating a 2D projection of the sample on a digital detector	Protein sample suspended in amorphous ice
4	In-cell NMR	Same as NMR but used to study proteins inside living cells	Labelled proteins samples
5	Cryo-EM tomography	3D reconstruction using tomography	Protein sample suspended in amorphous ice

ture of proteins. When a solution of labelled protein is placed in a magnetic field and subjected to different radio frequencies, then there is a change in the resonance of different atoms in the proteins. In an externally applied magnetic field, such atoms can flip between two states, viz. against or aligned with the magnetic field. So when the atoms are aligned against the external magnetic field the energy state of the atom is higher and this energy is a function of the rate at which the atoms resonate. This resonation is used to interpret and deduce the structure of the protein. On the basis of atoms selected for labelling, NMR spectroscopy is commonly of two types: (1) 1H NMR (to determine the type and number of H atoms in molecule) and (2) 13C NMR (determine the type of carbon atom in the molecule). NMR spectroscopy is better to determine the structure of proteins in the size ranging from 5 to 25 kDa by identifying carbon-hydrogen frameworks within molecules.

4.3.3.3 Cryo-Electron Microscopy

Electrons when accelerated in vacuum are 100,000 times shorter in wavelength than visible range, which makes it possible to resolve the points of few hundred nanometers apart. TEM technique uses this principle and has become a versatile tool in studying the protein structure at cryogenic temperature. Cryo-electron microscopy allows the observation of specimens in their native environment unlike X-ray crystallography. A thin film of a sample, aqueous solution, is rapidly frozen on a support grid and then placed in the high vacuum, where it is cooled with liquid nitrogen. Projection images of multiple copies of the molecule in random orientations are recorded, and 3D reconstructions of these images are performed using cryo-electron tomography. Transmission electron cryo microscopy was successful in determining the macromolecular structure considered too complex or large to be resolved by NMR or XRD [89]. The first protein structure to be solved using electron microscopy was bacteriorhodopsin [90–92]. Structures at near atomic resolution of viruses, ribosomes, mitochondria, and enzyme complexes had been determined using cryo-electron microscopy [93]. A recent report on complex structure between *E. coli* β-galactosidase and inhibitor phenylethyl b-D-thiogalactopyranoside (PETG) is determined at ~2.2 angstroms (Å) [94].

4.3.4 Structural Proteomics Study and Pathway

The vast amount of data generated from human genome project has provided vast opportunity to work on BIG data and omics. The translation of sequence information at protein level and further understanding of the molecular and functional aspects of protein has paved a way to understand the concept of "structural proteomics" or "structural genomics", the determination of 3D structure of protein on a genome-wide scale.

There is a rise in the use of high-throughput methods for protein production, structure determination, and functional analysis in order to scrutinize the growing protein universe and use it for translational research. The model organisms used for study of whole proteome till date are *Thermotogo maritima*, *Mycobacterium tuberculosis*, *Methanobacterium thermoautotrophicum*, and other *Archaebacteria*. Figure 4.2 gives a brief overview of pathways followed for structure determination using *in silico* and *in vitro* approaches. A structural proteomics study of the archaeon *Methanobacterium thermoautotrophicum* on a set of 424 non-membrane proteins was performed. These proteins were cloned, expressed, and structurally characterized. Out of 24 crystallized proteins, only 11 were diffracted for appropriate resolution. Furthermore, in NMR spectra, out of 100 soluble proteins tested, only 33 gave excellent spectra that could be used for structural determination. Similar work on *Methanobacterium thermoautotrophicum* was also performed by Yee et al. [95, 96]. Structural genomics study on thermophilic bacterium *Thermotoga maritime* was also attempted. 1376 of 1877 genes were cloned and attempted for expres-

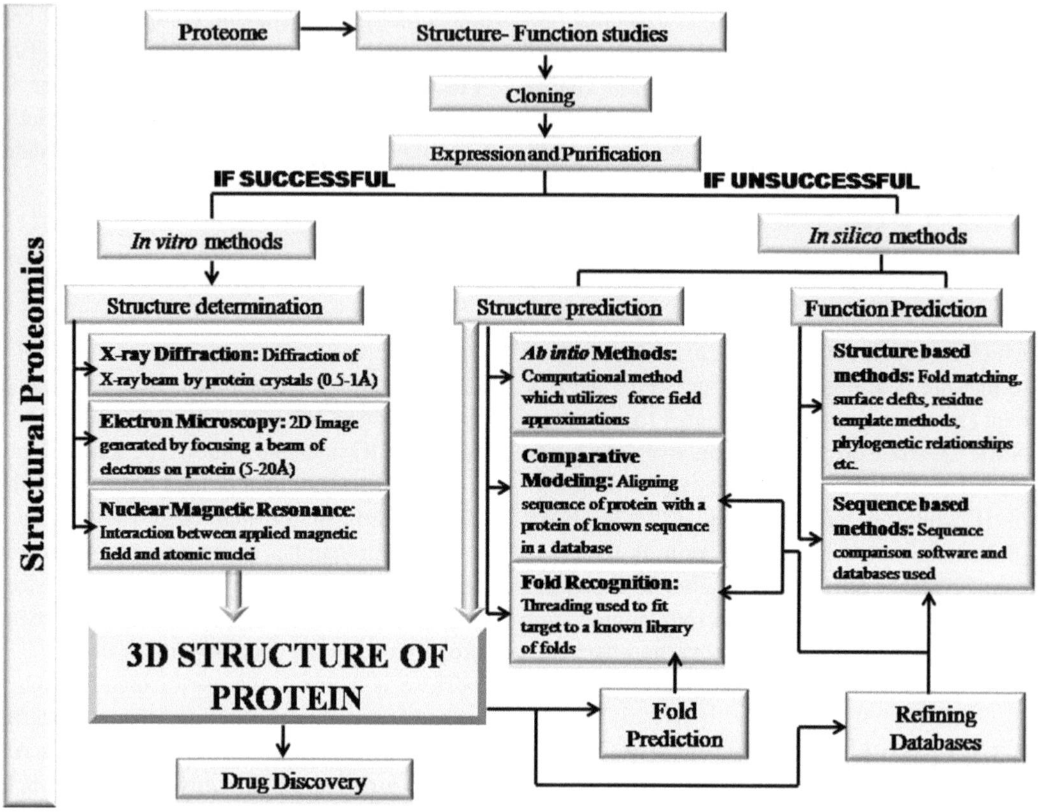

Fig. 4.2 Structural proteomics. Flow sheet represents method used for structure prediction

sion and purification. Crystallization condition for 432 proteins (23 %) of the *T. maritime proteome* was determined [97]. Structural proteomics study on uncharacterized proteins expressed in mouse macrophage cells to identify new drug targets for chronic obstructive pulmonary disease and arthritis was performed. Of the 318 macrophage gene processed, 220 of these were successfully cloned in bacterial expression system. 52 of these were soluble mouse macrophage proteins, however, structure of six carboxypeptidase inhibitor and acyl-CoA thioesterase were determined [98].

4.3.5 Structural Genomics Centre and Overview

Omics is helping to understand the holistic view to address the issues responsible for disease and understand the complex biological system. Structureomics study channelizes its efforts in determination of the 3D structure of protein and method development in making the entire process rapid and cost effective. Various consortium and structural genomics projects have been initiated by the Protein Structure Initiative in 2000 [99]. A brief overview of the different Structural Genomics Center and their roles in structural genomics project is highlighted (Table 4.5). Structural Genomics Centers could solve the structures of ~2800 proteins [100]. The information available from these consortia has allowed for accurate prediction of overall folds, to nearly 50 % of all known proteins, which is a significant increase from the past decade [101]. Amongst the many consortia formed, a few specifically targeted proteins related to infectious diseases. One of these is the TB consortium which focuses on structurally characterizing *M. tuberculosis* proteins. 250 novel proteins struc-

Table 4.5 The table lists various consortia present that perform different role in optimising different aspects of high through put methods in drug discovery

Structural genomics centre	# Structures reported	Expertise
RIKENStructural Genomics/Proteomics initiative	2743	Elucidation of protein functional networks via protein structural analysis
Joint Centre for Structural Genomics	1602	Focuses on the human microbiome
Structural Genomics Consortium	1386	Improve crystal formation by reductive methylation and limited proteolysis
New York Structural Genomics Research Consortium	1041	Focuses on industrialised protein production and structure determination followed by functional annotation and dissemination
Centre for Structural Genomics of Infectious Diseases	795	Determining structures of proteins/molecules that are involved in pathogenesis and infection in humans
TB Structural Genomics Consortium	285	Determination and analysis of structures of proteins from *Mycobacterium tuberculosis*
Centre for Eukaryotic Structural Genomics	218	Use cell free eukaryotic wheat germ extract for protein expression
Southeast Collaboratory for Structural Genomics	121	Focuses on development of high throughput structure determination methods
Structural Proteomics in Europe	119	Structure determination of biomedically relevant targets
Berkeley Structural Genomics Centre	101	Focuses on determining protein structures of two organisms- Mycoplasma genitalium and Mycoplasma pneumoniae.
Structural Genomics of Pathogenic Protozoa Consortium	71	Structure determination of proteins from trypanosomatid and malarial parasites using co-crystallisation and fragment cocktail crystallography
Enzyme Discovery for Natural Product Biosynthesis	63	Focuses on identification of new natural product pathways
New York Consortium on Membrane Protein Structure	57	Uses ultraviolet absorbance and light scattering to identify the best detergents for solubilisation of membrane proteins
Ontario Centre for Structural Proteomics	33	Use X-ray crystallography and NMR for structure determination

 # number

tures were solved by the consortia unlike previously reported eight structures from traditional method. These structures were useful in gaining functional insights about the protein and the mode of drug resistance [102]. The PSI program reported ~ 4000 unique structures into the Protein Data Bank (http://www.pdb.org/). The different PSI centers have contributed to 8 % of novel and 20 % of uncharacterized protein family structures [103]. The SGC project is responsible for one quarter of the total structural coverage of the human proteome available in the PDB [104].

Analysis of Protein Data Bank revealed expression organisms from prokaryotes to eukaryotes and acellular system are used for overproduction of proteins. The organism mostly preferred for protein expression are *Escherichia coli* and its different strains, *Spodoptera frugiperda*, *Tricho plusiani*, *Pichia pastoris*, *Saccharomyces cerevisiae*, *Cricetulus griseus*, *Drosophila melanogaster*, and cell-free synthesis. *Escherichia coli* is the most preferred of all for overexpression and this is probably because *E. coli* has high growth rate and low cost media and is non-pathogenic. Although reports do suggest that protein with posttranslational modification, large-sized proteins, and proteins that fold in the presence of folding machinery in the cell are not soluble in *E. coli* [105] are purified using eukaryotic system that includes yeast and mammalian cells.

The model organisms on which major work on structural genomics is focused on *Thermus thermophiles*, *Escherichia coli*, *Saccharomyces cerevisiae*, *Homo sapiens*, *Bos Taurus*, *Deinococcus radiodurans*, *Plasmodium falciparum 3D7*, *Drosophila melanogaster*, etc. The candidate organisms like *Thermus thermophiles* and *Escherichia coli* were initially studied in structural proteomics, because handling the organism with small size genome with less advanced technology was probably a feasible task [106]. These organisms are also known to share similarity with sequence and function of eukaryotic proteins, but are often smaller and more robust [95].

4.3.6 Advantages of Structural Proteomics

The era of structural genomics will make an immense impact on protein fold prediction, protein engineering, drug discovery, and basic and translational research. Structureomics will lead to revolutionary developments and automation in cloning, protein expression and purification, characterization, structure solution with NMR and crystallography, etc. Although omics study is not a "hypothesis-driven" research, it has the potential to answer certain key questions about biological function. The work on structureomics and extraction of information of sequence, structure, and function for application is based on certain assumptions like (1) proteins that have similar structures will mostly have similar functions; (2) structures of a protein can help in defining function of the molecule; and (3) functionally related proteins have conserved structures compared to sequences. Proteins with less than 10 % sequence similarity can still fold into similar structures, and in the absence of functional data, the fold of a protein can provide important clues about the function it may perform.

The development of high-throughput procedures will help determine several structures of proteins, protein-protein complexes, and protein-drug complexes that provide a knowledge base and different unknown aspects of structural biology. With the increase in simple protein structure, it is possible to identify novel folds, and with expanding databases, it will lead to accuracy in protein structure prediction. The atomic level detailed structure of protein does not have the ability to predict the conformational change in protein. Hence, there is a need to enhance our computational biophysics understanding to make accurate predictions about changes in macromolecular structures.

Structural genomics and association with functional genomics can help us to understand the structure and function of the proteins encoded by the novel genes. Knowledge of the structural details of proteins gives a clearer perspective of a protein to be an effective drug target. It allows for selection of molecules with minimum side effects and helps in optimization of the lead molecule. This makes them better candidates for entering a clinical trial, which can lead to discovery of a new drug [107].

Structureomics study will also generate prospects for method-oriented structural biologists as ample amount of "difficult" X-ray data sets and NMR spectra is produced. Thousands of clones and expression systems prepared during structural studies can be a wealth for specific in-depth biochemical studies. Structural genomics study on enzymes will provide detailed mechanism of catalysis of enzymes [108]. The presence of large number of structures of thermostable proteins will aid in engineering of industrial enzymes. Structureomics information of pathogenic organism will provide prospects for structure-based drug design, high-throughput screening, and combinatorial chemistry approaches. Accumulation of large amount of data in coming years may provide a system for structure-based computational toxicology study.

The information about structural details of the proteome will have an immediate boost on medicinal chemistry and molecular pharmacology. It also has an increasing impact on disciplines such as neurobiology, developmental biology, immunology, and molecular medicine.

4.3.7 Shortcomings of Structureomics

Enormous amount of fund and efforts have been applied in understanding the omics. Various consortiums are being developed to deal with different bottleneck existing in the pathway of high-throughput screening. The high-throughput approach may not reveal the complexity in structural biology as the (1) expression and purification of large complicated proteins is not possible and is challenging in the present scenario; (2) various others problems of yield, solubility, pseudosymmetry, and crystal twinning that exist will also appear in high-throughput approach. The conformational changes in the protein, different modes of aggregation, and precipitation will also influence the high-throughput approach [108]. Intrinsically disordered proteins break the paradigm of structure function correlation. Study on intrinsically disordered protein has revealed the fact that such proteins acquire ordered structure only when bound to it interacting partner. Such intrinsically disordered proteins are hurdles to structure-based drug discovery.

4.4 Summary

Understanding the shape that a protein molecule adopts to perform various functions in the cell is necessary to regulate these molecules. SBDD exploits this structure information for designing small molecule inhibitor to alter the activity of the target molecule. It also facilitates targeting a molecule that is important to design an inhibitor that is highly specific in nature which is a fundamental prerequisite for successful treatment. The information available from traditional structural biology methods has a lacuna that needs to be addressed. However, recent advancement in the field of structureomics has paved a way to successful determination of multiple structures and also in widening the bottlenecks to have a clear picture of protein at structure level. Thousands of different structures are deposited by various structure biology consortia which will not only enhance the knowledge of structural biology but also be useful in drug discovery and translational research.

References

1. Banaszak LJ (2000) Foundations of structural biology. Academic, San Diego
2. Goodsell DS (2011) Atomic evidence: the foundations of structural molecular biology. Sci Prog 94:414–430
3. Hughes JP, Rees S, Kalindjian SB, Philpott KL (2011) Principles of early drug discovery. Br J Pharmacol 162:1239–1249
4. Kurosawa G, Akahori Y, Morita M, Sumitomo M, Sato N, Muramatsu C, Eguchi K, Matsuda K, Takasaki A, Tanaka M, Iba Y, Hamada-Tsutsumi S, Ukai Y, Shiraishi M, Suzuki K, Kurosawa M, Fujiyama S, Takahashi N, Kato R, Mizoguchi Y, Shamoto M, Tsuda H, Sugiura M, Hattori Y, Miyakawa S, Shiroki R, Hoshinaga K, Hayashi N, Sugioka A, Kurosawa Y (2008) Comprehensive screening for antigens overexpressed on carcinomas via isolation of human mAbs that may be therapeutic. Proc Natl Acad Sci U S A 105:7287–7292
5. Zanders ED, Bailey DS, Dean PM (2002) Probes for chemical genomics by design. Drug Discov Today 7:711–718
6. Lipinski CA (2004) Lead- and drug-like compounds: the rule-of-five revolution. Drug Discov Today Technol 1:337–341
7. Golebiowski A, Klopfenstein SR, Portlock DE (2003) Lead compounds discovered from libraries: part 2. Curr Opin Chem Biol 7:308–325
8. Lipinski CA, Lombardo F, Dominy BW, Feeney PJ (2012) Experimental and computational approaches to estimate solubility and permeability in drug discovery and development settings. Adv Drug Deliv Rev 64:4–17
9. Hajduk PJ, Galloway WR, Spring DR (2011) Drug discovery: a question of library design. Nature 470:42–43
10. Verdonk ML, Hartshorn MJ (2004) Structure-guided fragment screening for lead discovery. Curr Opin Drug Discov Devel 7:404–410
11. Campbell SF (2000) Science, art and drug discovery: a personal perspective. Clin Sci (Lond) 99:255–260

12. Chou JJ, Kaufman JD, Stahl SJ, Wingfield PT, Bax A (2002) Micelle-induced curvature in a water-insoluble HIV-1 Env peptide revealed by NMR dipolar coupling measurement in stretched polyacrylamide gel. J Am Chem Soc 124:2450–2451
13. Stoll V, Qin W, Stewart KD, Jakob C, Park C, Walter K, Simmer RL, Helfrich R, Bussiere D, Kao J, Kempf D, Sham HL, Norbeck DW (2002) X-ray crystallographic structure of ABT-378 (lopinavir) bound to HIV-1 protease. Bioorg Med Chem 10:2803–2806
14. Davis AM, Teague SJ, Kleywegt GJ (2003) Application and limitations of X-ray crystallographic data in structure-based ligand and drug design. Angew Chem Int Ed Engl 42:2718–2736
15. Shoichet B, Bussiere D (2000) Macromolecular crystallography and lead discovery: possibilities and limitations. J Mol Biol 295:337–356
16. Bollag G, Tsai J, Zhang J, Zhang C, Ibrahim P, Nolop K, Hirth P (2012) Vemurafenib: the first drug approved for BRAF-mutant cancer. Nat Rev Drug Discov 11:873–886
17. Lynch TJ, Bell DW, Sordella R, Gurubhagavatula S, Okimoto RA, Brannigan BW, Harris PL, Haserlat SM, Supko JG, Haluska FG, Louis DN, Christiani DC, Settleman J, Haber DA (2004) Activating mutations in the epidermal growth factor receptor underlying responsiveness of non — small-cell lung cancer to gefitinib. N Engl J Med 350:2129–2139
18. Miller M, Schneider J, Sathyanarayana BK, Toth MV, Marshall GR, Clawson L, Selk L, Kent SB, Wlodawer A (1989) Structure of complex of synthetic HIV-1 protease with a substrate-based inhibitor at 2.3 A resolution. Science 246:1149–1152
19. Lapatto R, Blundell T, Hemmings A, Overington J, Wilderspin A, Wood S, Merson JR, Whittle PJ, Danley DE, Geoghegan KF (1989) X-ray analysis of HIV-1 proteinase at 2.7 a resolution confirms structural homology among retroviral enzymes. Nature 342:299–302
20. Wong S, Witte ON (2004) The BCR-ABL story: bench to bedside and back. Annu Rev Immunol 22:247–306
21. Wood ER, Truesdale AT, McDonald OB, Yuan D, Hassell A, Dickerson SH, Ellis B, Pennisi C, Horne E, Lackey K, Alligood KJ, Rusnak DW, Gilmer TM, Shewchuk L (2004) A unique structure for epidermal growth factor receptor bound to GW572016 (Lapatinib): relationships among protein conformation, inhibitor off-rate, and receptor activity in tumor cells. Cancer Res 64:6652–6659
22. Dufe VT, Qiu W, Müller IB, Hui R, Walter RD, Al-Karadaghi S (2007) Crystal structure of plasmodium falciparum spermidine synthase in complex with the substrate decarboxylated S-adenosylmethionine and the potent inhibitors 4MCHA and AdoDATO. J Mol Biol 373:167–177
23. Varghese JN (1999) Development of neuraminidase inhibitors as anti-influenza virus drugs. Drug Dev Res 46:176–196
24. Smaill JB, Rewcastle GW, Loo JA, Greis KD, Chan OH, Reyner EL, Lipka E, Showalter HH, Vincent PW, Elliott WL (2000) Tyrosine kinase inhibitors. 17. Irreversible inhibitors of the epidermal growth factor receptor: 4-(phenylamino) quinazoline-and 4-(phenylamino) pyrido [3, 2-d] pyrimidine-6-acrylamides bearing additional solubilizing functions. J Med Chem 43:1380–1397
25. Cario H, Smith DE, Blom H, Blau N, Bode H, Holzmann K, Pannicke U, Hopfner K-P, Rump E-M, Ayric Z (2011) Dihydrofolate reductase deficiency due to a homozygous DHFR mutation causes megaloblastic anemia and cerebral folate deficiency leading to severe neurologic disease. Am J Hum Genet 88:226–231
26. Schweitzer BI, Dicker AP, Bertino JR (1990) Dihydrofolate reductase as a therapeutic target. FASEB J 4:2441–2452
27. Sharma M, Chauhan PM (2012) Dihydrofolate reductase as a therapeutic target for infectious diseases: opportunities and challenges. Future med chem 4:1335–1365
28. Patick A, Binford S, Brothers M, Jackson R, Ford C, Diem M, Maldonado F, Dragovich P, Zhou R, Prins T (1999) In vitro antiviral activity of AG7088, a potent inhibitor of human rhinovirus 3C protease. Antimicrob Agents Chemother 43:2444–2450
29. Masuda Y, Karasawa T (1993) Inhibitory effect of zonisamide on human carbonic anhydrase in vitro. Arzneimittelforschung 43:416–418
30. Bissett D, O'Byrne KJ, Von Pawel J, Gatzemeier U, Price A, Nicolson M, Mercier R, Mazabel E, Penning C, Zhang MH (2005) Phase III study of matrix metalloproteinase inhibitor prinomastat in non–small-cell lung cancer. J Clin Oncol 23:842–849
31. Van Zandt MC, Jones ML, Gunn DE, Geraci LS, Jones JH, Sawicki DR, Sredy J, Jacot JL, DiCioccio AT, Petrova T (2005) Discovery of 3-[(4, 5, 7-trifluorobenzothiazol-2-yl) methyl] indole-N-acetic acid (lidorestat) and congeners as highly potent and selective inhibitors of aldose reductase for treatment of chronic diabetic complications. J Med Chem 48:3141–3152
32. McPherson A (2004) Protein crystallization in the structural genomics era. J Struct Funct Genom 5:3–12
33. Dale GE, Oefner C, D'Arcy A (2003) The protein as a variable in protein crystallization. J Struct Biol 142:88–97
34. Acharya KR, Lloyd MD (2005) The advantages and limitations of protein crystal structures. Trends Pharmacol Sci 26:10–14
35. Anderson AC (2003) The process of structure-based drug design. Chem Biol 10:787–797
36. Patil R, Das S, Stanley A, Yadav L, Sudhakar A, Varma AK (2010) Optimized hydrophobic interactions and hydrogen bonding at the target-ligand interface leads the pathways of drug-designing. PLoS One 5:e12029
37. McGovern SL, Caselli E, Grigorieff N, Shoichet BK (2002) A common mechanism underlying promiscuous

inhibitors from virtual and high-throughput screening. J Med Chem 45:1712–1722

38. Joseph-McCarthy D (2009) Challenges of fragment screening. J Comput Aided Mol Des 23:449–451
39. Song Y, Chen W, Kang D, Zhang Q, Zhan P, Liu X (2014) "Old friends in new guise": exploiting privileged structures for scaffold re-evolution/refining. Comb Chem High Throughput Screen 17:536–553
40. Chen H, Zhou X, Wang A, Zheng Y, Gao Y, Zhou J (2015) Evolutions in fragment-based drug design: the deconstruction–reconstruction approach. Drug Discov Today 20:105–113
41. Song Y, Zhan P, Liu X (2013) Heterocycle-thioacetic acid motif: a privileged molecular scaffold with potent, broad-ranging pharmacological activities. Curr Pharm Des 19:7141–7154
42. Liu Y, Zhou E, Yu K, Zhu J, Zhang Y, Xie X, Li J, Jiang H (2008) Discovery of a novel CCR5 antagonist lead compound through fragment assembly. Molecules 13:2426–2441
43. Viegas-Júnior C, Danuello A, da Silva Bolzani V, Barreiro EJ, Fraga CAM (2007) Molecular hybridization: a useful tool in the design of new drug prototypes. Curr Med Chem 14:1829–1852
44. Manetsch R, Krasinski A, Radic Z, Raushel J, Taylor P, Sharpless KB, Kolb HC (2004) In situ click chemistry: enzyme inhibitors made to their own specifications. J Am Chem Soc 126:12809–12818
45. Thirumurugan P, Matosiuk D, Jozwiak K (2013) Click chemistry for drug development and diverse chemical–biology applications. Chem Rev 113:4905–4979
46. Brik A, Wu C-Y, Wong C-H (2006) Microtiter plate based chemistry and in situ screening: a useful approach for rapid inhibitor discovery. Org biomol chem 4:1446–1457
47. Hubbard RE (2008) Fragment approaches in structure-based drug discovery. J Synchrotron Radiat 15:227–230
48. Hajduk PJ, Greer J (2007) A decade of fragment-based drug design: strategic advances and lessons learned. Nat Rev Drug Discov 6:211–219
49. Ji H, Zhang W, Zhang M, Kudo M, Aoyama Y, Yoshida Y, Sheng C, Song Y, Yang S, Zhou Y, Lu J, Zhu J (2003) Structure-based de novo design, synthesis, and biological evaluation of non-azole inhibitors specific for lanosterol 14alpha-demethylase of fungi. J Med Chem 46:474–485
50. Honma T, Hayashi K, Aoyama T, Hashimoto N, Machida T, Fukasawa K, Iwama T, Ikeura C, Ikuta M, Suzuki-Takahashi I, Iwasawa Y, Hayama T, Nishimura S, Morishima H (2001) Structure-based generation of a new class of potent Cdk4 inhibitors: new de novo design strategy and library design. J Med Chem 44:4615–4627
51. Schmidt JM, Mercure J, Tremblay GB, Page M, Kalbakji A, Feher M, Dunn-Dufault R, Peter MG, Redden PR (2003) De novo design, synthesis, and evaluation of novel nonsteroidal phenanthrene ligands for the estrogen receptor. J Med Chem 46:1408–1418
52. Pierce AC, Rao G, Bemis GW (2004) BREED: Generating novel inhibitors through hybridization of known ligands. Application to CDK2, p38, and HIV protease. J Med Chem 47:2768–2775
53. Lloyd DG, Buenemann CL, Todorov NP, Manallack DT, Dean PM (2004) Scaffold hopping in de novo design. Ligand generation in the absence of receptor information. J Med Chem 47:493–496
54. Rich K (2004) An overview of clinical trials. J Vasc Nurs 22:32–34
55. Liang BC (2002) The drug development process III: phase IV clinical trials. Hosp Physician 38:42
56. Sims J, Miracle VA (2002) Phases of a clinical trial. Dimens Crit Care Nurs 21:152–153
57. Kim SH (1998) Shining a light on structural genomics. Nat Struct Biol 5(Suppl):643–645
58. Overington JP, Al-Lazikani B, Hopkins AL (2006) How many drug targets are there? Nat Rev Drug Discov 5:993–996
59. Guss JM, Merritt EA, Phizackerley RP, Hedman B, Murata M, Hodgson KO, Freeman HC (1988) Phase determination by multiple-wavelength x-ray diffraction: crystal structure of a basic "blue" copper protein from cucumbers. Science 241:806–811
60. Abola E, Kuhn P, Earnest T, Stevens RC (2000) Automation of X-ray crystallography. Nat Struct Biol 7(Suppl):973–977
61. Muchmore SW, Olson J, Jones R, Pan J, Blum M, Greer J, Merrick SM, Magdalinos P, Nienaber VL (2000) Automated crystal mounting and data collec-tion for protein crystallography. Structure 8:R243–R246
62. Adams PD, Grosse-Kunstleve RW (2000) Recent developments in software for the automation of crystallographic macromolecular structure determination. Curr Opin Struct Biol 10:564–568
63. Karve TM, Cheema AK (2011) Small changes huge impact: the role of protein posttranslational modifications in cellular homeostasis and disease. J Amino Acids 2011:207691
64. Benson DA, Clark K, Karsch-Mizrachi I, Lipman DJ, Ostell J, Sayers EW (2014) GenBank. Nucleic Acids Res 42:D32–D37
65. Larkin MA, Blackshields G, Brown NP, Chenna R, McGettigan PA, McWilliam H, Valentin F, Wallace IM, Wilm A, Lopez R, Thompson JD, Gibson TJ, Higgins DG (2007) Clustal W and Clustal X version 2.0. Bioinformatics 23:2947–2948
66. Altschul SF, Madden TL, Schaffer AA, Zhang J, Zhang Z, Miller W, Lipman DJ (1997) Gapped BLAST and PSI-BLAST: a new generation of protein database search programs. Nucleic Acids Res 25:3389–3402
67. Bork P (2000) Powers and pitfalls in sequence analysis: the 70% hurdle. Genome Res 10:398–400
68. Baumann U, Wu S, Flaherty KM, McKay DB (1993) Three-dimensional structure of the alkaline protease

of Pseudomonas aeruginosa: a two-domain protein with a calcium binding parallel beta roll motif. EMBO J 12:3357–3364
69. Friedberg I (2006) Automated protein function prediction--the genomic challenge. Brief Bioinform 7:225–242
70. Finn RD, Bateman A, Clements J, Coggill P, Eberhardt RY, Eddy SR, Heger A, Hetherington K, Holm L, Mistry J, Sonnhammer EL, Tate J, Punta M (2014) Pfam: the protein families database. Nucleic Acids Res 42:D222–D230
71. Hulo N, Bairoch A, Bulliard V, Cerutti L, Cuche BA, de Castro E, Lachaize C, Langendijk-Genevaux PS, Sigrist CJ (2008) The 20 years of PROSITE. Nucleic Acids Res 36:D245–D249
72. Henikoff JG, Greene EA, Pietrokovski S, Henikoff S (2000) Increased coverage of protein families with the blocks database servers. Nucleic Acids Res 28:228–230
73. Attwood TK, Coletta A, Muirhead G, Pavlopoulou A, Philippou PB, Popov I, Roma-Mateo C, Theodosiou A, Mitchell AL (2012) The PRINTS database: a fine-grained protein sequence annotation and analysis resource–its status in 2012, Database (Oxford) 2012 bas019.
74. Norin M, Sundstrom M (2002) Structural proteomics: developments in structure-to-function predictions. Trends Biotechnol 20:79–84
75. Arnold K, Bordoli L, Kopp J, Schwede T (2006) The SWISS-MODEL workspace: a web-based environment for protein structure homology modelling. Bioinformatics 22:195–201
76. Hardin C, Pogorelov TV, Luthey-Schulten Z (2002) Ab initio protein structure prediction. Curr Opin Struct Biol 12:176–181
77. Watson JD, Laskowski RA, Thornton JM (2005) Predicting protein function from sequence and structural data. Curr Opin Struct Biol 15:275–284
78. Holm L, Sander C (1993) Protein structure comparison by alignment of distance matrices. J Mol Biol 233:123–138
79. Krissinel E, Henrick K (2004) Secondary-structure matching (SSM), a new tool for fast protein structure alignment in three dimensions. Acta Crystallogr D Biol Crystallogr 60:2256–2268
80. Madej T, Gibrat JF, Bryant SH (1995) Threading a database of protein cores. Proteins 23:356–369
81. Binkowski TA, Freeman P, Liang J (2004) pvSOAR: detecting similar surface patterns of pocket and void surfaces of amino acid residues on proteins. Nucleic Acids Res 32:W555–W558
82. Ferre F, Ausiello G, Zanzoni A, Helmer-Citterich M (2004) SURFACE: a database of protein surface regions for functional annotation. Nucleic Acids Res 32:D240–D244
83. Tsuchiya Y, Kinoshita K, Nakamura H (2005) PreDs: a server for predicting dsDNA-binding site on protein molecular surfaces. Bioinformatics 21:1721–1723
84. Tuszynska I, Magnus M, Jonak K, Dawson W, Bujnicki JM (2015) NPDock: a web server for protein-nucleic acid docking. Nucleic Acids Res 43:W425–W430
85. Khandelia P, Yap K, Makeyev EV (2011) Streamlined platform for short hairpin RNA interference and transgenesis in cultured mammalian cells. Proc Natl Acad Sci U S A 108:12799–12804
86. Murphy GE, Jensen GJ (2007) Electron cryotomography. Biotechniques 43:413, 415, 417 passim
87. Berman HM, Westbrook J, Feng Z, Gilliland G, Bhat TN, Weissig H, Shindyalov IN, Bourne PE (2000) The protein data bank. Nucleic Acids Res 28:235–242
88. Berman HM, Bhat TN, Bourne PE, Feng Z, Gilliland G, Weissig H, Westbrook J (2000) The protein data bank and the challenge of structural genomics. Nat Struct Biol 7(Suppl):957–959
89. Jonic S, Venien-Bryan C (2009) Protein structure determination by electron cryo-microscopy. Curr Opin Pharmacol 9:636–642
90. Henderson R, Baldwin JM, Ceska TA, Zemlin F, Beckmann E, Downing KH (1990) Model for the structure of bacteriorhodopsin based on high-resolution electron cryo-microscopy. J Mol Biol 213:899–929
91. Kuhlbrandt W, Wang DN, Fujiyoshi Y (1994) Atomic model of plant light-harvesting complex by electron crystallography. Nature 367:614–621
92. Nogales E, Wolf SG, Downing KH (1998) Structure of the alpha beta tubulin dimer by electron crystallography. Nature 391:199–203
93. Kuhlbrandt W (2014) Cryo-EM enters a new era. Elife 3:e03678
94. Bartesaghi A, Merk A, Banerjee S, Matthies D, Wu X, Milne JL, Subramaniam S (2015) 2.2 Å resolution cryo-EM structure of β-galactosidase in complex with a cell-permeant inhibitor. Science 348:1147–51
95. Christendat D, Yee A, Dharamsi A, Kluger Y, Savchenko A, Cort JR, Booth V, Mackereth CD, Saridakis V, Ekiel I, Kozlov G, Maxwell KL, Wu N, McIntosh LP, Gehring K, Kennedy MA, Davidson AR, Pai EF, Gerstein M, Edwards AM, Arrowsmith CH (2000) Structural proteomics of an archaeon. Nat Struct Biol 7:903–909
96. Yee A, Pardee K, Christendat D, Savchenko A, Edwards AM, Arrowsmith CH (2003) Structural proteomics: toward high-throughput structural biology as a tool in functional genomics. Acc Chem Res 36:183–189
97. Lesley SA, Kuhn P, Godzik A, Deacon AM, Mathews I, Kreusch A, Spraggon G, Klock HE, McMullan D, Shin T (2002) Structural genomics of the Thermotoga maritima proteome implemented in a high-throughput structure determination pipeline. Proc Natl Acad Sci 99:11664–11669
98. Puri M, Robin G, Cowieson N, Forwood JK, Listwan P, Hu SH, Guncar G, Huber T, Kellie S, Hume DA,

Kobe B, Martin JL (2006) Focusing in on structural genomics: the University of Queensland structural biology pipeline. Biomol Eng 23:281–289

99. Chandonia JM, Brenner SE (2006) The impact of structural genomics: expectations and outcomes. Science 311:347–351
100. Berman HM, Westbrook JD, Gabanyi MJ, Tao W, Shah R, Kouranov A, Schwede T, Arnold K, Kiefer F, Bordoli L, Kopp J, Podvinec M, Adams PD, Carter LG, Minor W, Nair R, La Baer J (2009) The protein structure initiative structural genomics knowledgebase. Nucleic Acids Res 37:D365–D368
101. Dessailly BH, Nair R, Jaroszewski L, Fajardo JE, Kouranov A, Lee D, Fiser A, Godzik A, Rost B, Orengo C (2009) PSI-2: structural genomics to cover protein domain family space. Structure 17:869–881
102. Baker EN (2007) Structural genomics as an approach towards understanding the biology of tuberculosis. J Struct Funct Genom 8:57–65
103. Nair R, Liu J, Soong T-T, Acton TB, Everett JK, Kouranov A, Fiser A, Godzik A, Jaroszewski L, Orengo C (2009) Structural genomics is the largest contributor of novel structural leverage. J Struct Funct Genom 10:181–191
104. Edwards A (2009) Large-scale structural biology of the human proteome. Annu Rev Biochem 78:541–568
105. Rosano GL, Ceccarelli EA (2014) Recombinant protein expression in Escherichia coli: advances and challenges. Front Microbiol 5:172
106. Smith DR, Doucette-Stamm LA, Deloughery C, Lee H, Dubois J, Aldredge T, Bashirzadeh R, Blakely D, Cook R, Gilbert K, Harrison D, Hoang L, Keagle P, Lumm W, Pothier B, Qiu D, Spadafora R, Vicaire R, Wang Y, Wierzbowski J, Gibson R, Jiwani N, Caruso A, Bush D, Reeve JN et al (1997) Complete genome sequence of Methanobacterium thermoautotrophicum deltaH: functional analysis and comparative genomics. J Bacteriol 179:7135–7155
107. Russell RB, Eggleston DS (2000) New roles for structure in biology and drug discovery. Nat Struct Biol 7(Suppl):928–930
108. Hol WG (2000) Structural genomics for science and society. Nat Struct Biol 7(Suppl):964–966

Biosensors for Metabolic Engineering

5

Qiang Yan and Stephen S. Fong

Abbreviations

2-PS	2-Pyrone synthase
acetyl-CoA	Acetyl coenzyme A
ATF	Artificial transcription factor
BLA	β-Lactamase
bPBP	Bacterial periplasmic binding protein
CIDs	Chemical inducers of dimerization
DBDs	DNA-binding domains
ER	Estrogen receptor
eyfp	Enhanced yellow fluorescence protein
FACS	Fluorescence-activated cell sorting
FAEE	Fatty acid ethyl ester
GC	Gas chromatography
GFP	Green fluorescent protein
HHRs	Hammerhead ribozymes
HMG-CoA	Hydroxymethylglutaryl-CoA
HPLC	High-performance liquid chromatography
IPP	Isopentenyl pyrophosphate
IPTG	β-D-1-Thiogalactopyranoside
LBD	Ligand-binding domain
MAGE	Multiplex automated genome engineering
MBP	Maltose-binding protein
MRTF	Metabolite-responsive transcription factor
RBS	Ribosome binding site
RD	Regulatory domain
RFP	Red fluorescent protein
SDS	Sodium dodecyl sulfate
TAL	Triacetic acid lactone
TATB	1,3,5-Triamino-2,4,6-trinitrobenzene
TPP	Thiamine pyrophosphate

5.1 Introduction

Metabolic engineering broadly encompasses the engineering of biological systems to enable production of a wide variety of valuable compounds for chemicals including biofuels, pharmaceuticals, nutraceuticals, bulk chemicals, and materials [30, 76, 78]. To produce these value-added compounds, efficient biosynthesis pathways must be constructed in appropriate host. This often requires extensive optimization to reach economically viable titers, yields, and productivity. However, current approaches require a significant investment of time and resources for

Q. Yan
Department of Chemical and Life Science Engineering, Virginia Commonwealth University, West Hall, Room 422, 601 West Main Street, 843028, Richmond, VA 23284-3028, USA

S.S. Fong (✉)
Department of Chemical and Life Science Engineering, Virginia Commonwealth University, West Hall, Room 422, 601 West Main Street, 843028, Richmond, VA 23284-3028, USA

Center for the Study of Biological Complexity, Virginia Commonwealth University, Richmond, VA, USA
e-mail: ssfong@vcu.edu

S. Singh (ed.), *Systems Biology Application in Synthetic Biology*,
DOI 10.1007/978-81-322-2809-7_5

each individual pathway, limiting the number of compounds to which these strategies can be applied and thus the scalability of biosynthetic approaches [30]. Synthetic biology is a fast-growing field that develops new tools for biological engineering and can be applied as a means of interrogating pathway optimization in a rigorous, detailed manner. Synthetic biology has proven effective in the development of new tools and technologies that support the design, construction, and optimization of complex biological systems. As engineered microbial biosynthesis platforms have the most immediate practical applications in terms of development of industrial products, it is not surprising that many of the advances in tool development have been directed to metabolic pathway engineering [27, 28, 58]. Among these new tools, biosensors represent a significant contribution from synthetic biology and have been increasingly used in metabolic engineering. Here, we provide information and recent work on the development of metabolite biosensors and their applications for metabolic engineering.

One early definition of biosensors termed them as a device incorporating a biological sensing element either intimately connected to or integrated within a transducer [66]. The common aim is to produce a digital electronic signal which is proportional to the concentration of a specific chemical or set of chemicals [66]. It was proposed that enzymes could be immobilized in conjunction with electrochemical detectors to form "enzyme electrodes" which would expand the analytical range of the base sensor. After the first wave of initial biosensor design, more and more knowledge has been gained from natural biological systems (e.g., tissue, microorganism, organelles, enzymes, antibodies, nucleic acid, etc.) enabling improvements in biosensors. Biosensors have been widely used in various fields such as clinical applications, environment diagnostics, or food analysis. One common example of a commercial biosensor is the blood glucose biosensor, which uses the enzyme glucose oxidase to break blood glucose by oxidizing glucose to produce two electrons to reduce FAD (a component of the enzyme) to $FADH_2$. $FADH_2$ is then oxidized by an electrode as a method of measuring the glucose concentration [68].

For metabolic engineering applications, metabolite biosensors have been developed as genetically encoded proteins or RNA-based biosensors that interact with a metabolite to generate an actuator output [30, 35]. The output part of a metabolite biosensor generates detectable phenotypes through modulating transcription rates, translation rates, or protein activity to control protein expression or function. Over the past few decades, metabolite biosensors have been widely used to select high-producing strains in high-throughput screens, sensing of a desirable product in selective conditions, and dynamic control of metabolic flux.

Biosensors can be coupled to readable outputs such as fluorescence to semiquantitatively report the concentration of a target compound. This approach is frequently used for high-throughput screening of high-producing strains and features distinct advantages over conventional methods such as gas chromatography (GC) and high-performance liquid chromatography (HPLC) since (1) biosensor-mediated quantification avoids time-consuming sample preparation and has much higher throughput than conventional chromatographic techniques; (2) metabolite biosensors are more suitable for detecting labile and low abundant metabolites such as acyl-phosphate, acyl-diphosphate, aldehyde, and acyl-CoAs, which are difficult to measure accurately by conventional methods; and (3) metabolite biosensors allow real-time monitoring of metabolite dynamics in living cells, which is impossible to study using chromatographic methods. These reporter outputs may also help coordinate complementary perturbation of the culture environment itself (mixing, nutrient addition, time of harvest) to further improve production [43].

Second, biosensors can be engineered to couple the sensing of a desirable product or intermediate metabolite with a fitness advantage for the cell by expressing a gene necessary for survival under selective conditions [13, 45]. The difference in cell growth allows direct enrichment of fast-growing cells from mutant libraries, which allows an easy selection for desirable production characteristics.

Third, metabolite biosensors can also be used to control metabolic flux dynamically [12, 31, 77, 79]. The actuator can be designed to tune pathway enzyme expression or posttranslational parameters in response to the level of the relevant metabolite, allowing for dynamic control of a metabolic pathway, which not only reduces toxic intermediate accumulation but also saves carbon and energy that are otherwise diverted to synthesize unnecessary proteins or intermediates [30]. Overall, the emerging tools to engineer biosensors and their applications toward metabolic engineering have greatly advanced microbial production of a variety of chemicals.

This chapter will first discuss and classify metabolite biosensors into four categories based on their diverse mechanisms of sensing and functional output, including (1) transcription factor-based biosensor, (2) RNA-based sensors, (3) protein activity-based sensors, and (4) whole cell sensors. Next, we will present specific targets of biosensor applications in metabolic engineering. Then, we discuss tools for developing and designing metabolite biosensors. Finally, we discuss future directions of metabolite biosensors in the field of metabolic engineering and synthetic biology.

5.2 Types of Metabolite Biosensors

5.2.1 Transcription Factor-Based Biosensors

In nature, transcription factors are proteins involved in regulating gene expression by specific binding to chromosomal DNA, blocking or initiating transcription. For example, among 230 transcription factors in *Escherichia coli*, two of the well-studied examples are LacI and AraC. LacI is a transcription factor protein that control *lac* operon gene expression by lactose or its analogue, isopropyl β-D-1-thiogalactopyranoside (IPTG) [7, 21, 52]. In the absence of lactose, LacI binds specifically to the major groove of the operator region of the *lac* operon, resulting in the halt of RNA polymerase "read through." In the presence of lactose or IPTG, the small molecule binds to LacI, resulting in an allosteric change of its shape, subsequently causing an inability to bind to its target. Another example of a transcription factor was found in L-arabinose operon, the AraC, which regulates gene expression of *araA*, *araB*, and *araD* with or without arabinose. When arabinose is absent, the dimer AraC represses the expression of *araABD* by binding to araI1 and araO2 to form a loop. The loop prevents RNA polymerase from binding to the promoter of the ara operon, thereby blocking transcription. When arabinose is present, arabinose binds AraC and prevents AraC from forming of the DNA loop, thereby allowing transcription to proceed.

By using the ability of binding to small molecules such as sugars, sugar phosphates, amino acids, and lipids, natural metabolite-responsive transcription factors (MRTF) could be engineered as biosensors for metabolic engineering applications [13, 30, 35]. Typically, metabolite-responsive promoters with tunable output dynamic ranges can be engineered by inserting the cognate operator of a MRTF into a synthetic promoter to regulate gene expression (Fig. 5.1). Depending on the type of metabolite, two strategies can be implemented. One strategy is to integrate the cognate operator of a MRTF into a natural or synthetic promoter to regulate genes of interests. This type of strategy is suitable for intermediate/precursor metabolites such as acyl-CoA, malonyl-CoA [31, 74], and acetyl-CoA [77]. Since the intermediates are essential for both growth and chemical production, they are typically hard to monitor, and intracellular concentrations are expected to be moderate. Overexpression of downstream pathway genes usually results in unnecessary production of proteins and resources, which could adversely affect cell growth, while low expression of downstream genes usually is not able to obtain desired yields. According to this strategy, a variety of metabolite-actuated biosensors have been developed, such as FadR response to acyl-CoA for fatty acid ethyl

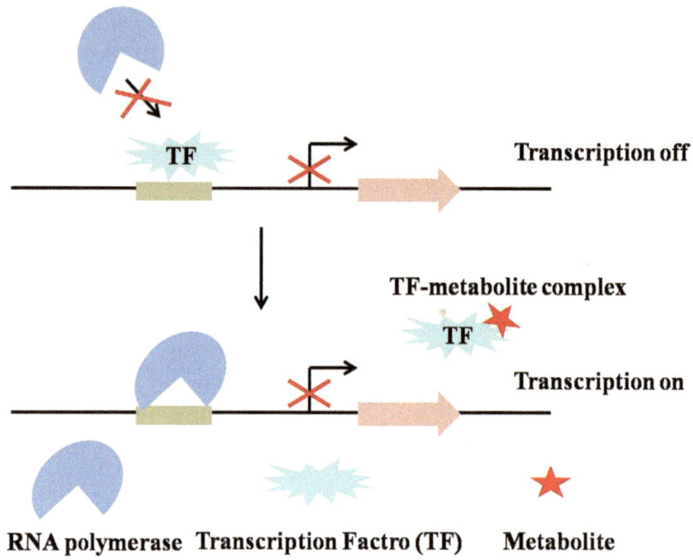

Fig. 5.1 Transcription factor-based metabolite biosensors

ester (FAEE) production and FapR response to malonyl-CoA for fatty acid biosynthesis [30, 73].

The second strategy is to screen for high-producing strains from a library of natural or engineered strains by using MRTF. This approach becomes particularly powerful when coupled with fluorescence-activated cell sorting (FACS). First of all, a natural MRTF-based biosensor is selected as a target, which usually shares similar structure to the desired metabolites. Then, various protein engineering methodologies (rational design or directed evolution, see discussion below) are utilized to alter the specificity of the MRTF to detect the target metabolite for which no natural sensor exists. By coupling a fluorescence protein under the control of recognized and regulated promoters, active variants could be rapidly selected. For example, AraC has been developed to sense arabinose structural analogues, such as D-arabinose [61], fructose, ribose [33, 54], and mevalonate [11].

5.2.2 RNA-Based Sensors

5.2.2.1 Transcription-Based RNA Sensors

Transcription-based RNA sensors are usually built upon aptamer domains to either facilitate or disrupt the formation of a terminator, which prevents the synthesis of long mRNAs, creating transcriptional repression or activation (Fig. 5.2a). Such engineered RNA sensors are usually only specific to limited metabolites, such as folinic acid and theophylline due to limited types of available aptamers [65, 70]. The screening output accuracy could be improved by increasing the copy numbers of the same riboswitch to a single transcription unit [70].

5.2.2.2 Translation-Based RNA Sensors

Riboswitches can be engineered to sense metabolites and regulate the secondary structure of mRNAs to either promote or inhibit the ribosome binding site (RBS) sequence from interacting with the ribosome, a strategy predominantly used by prokaryotes to modulate translation initiation (Fig. 5.2b). Synthetic riboswitches were engineered to sense various metabolites, such as theophylline [64], ammeline [14], and thiamine pyrophosphate [37]. For example, the *Escherichia coli* thiamine pyrophosphate (TPP) riboswitch was synthesized and cloned in front of a reporter *gfp* gene (encoding the green fluorescent protein, GFP) under the control of the plastid ribosomal operon promoter Prrn. A Shine-Dalgarno structure was designed in the riboswitch to confer

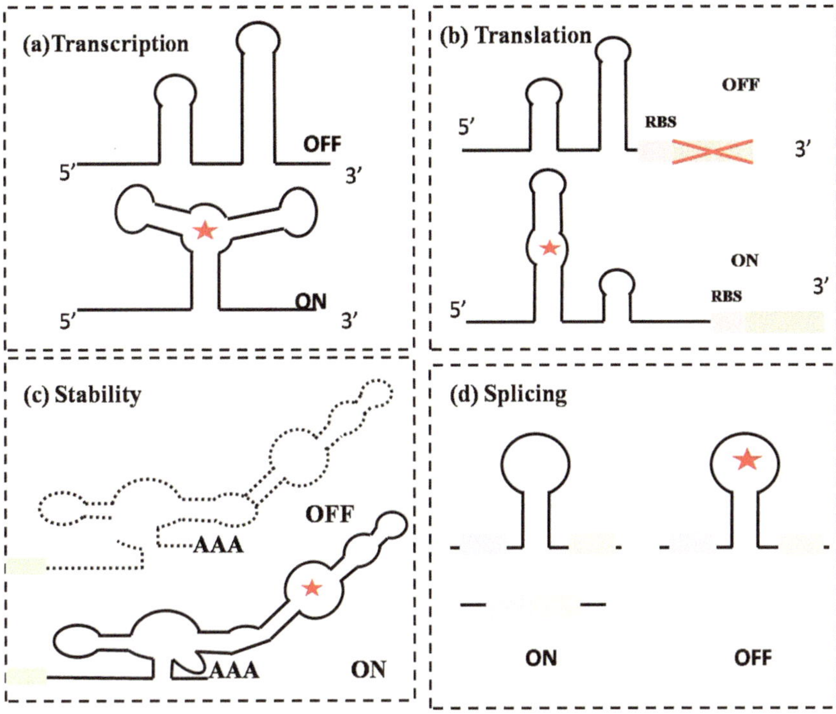

Fig. 5.2 RNA-based metabolite biosensors: (**a**) RNA-based metabolite biosensors control transcription. When metabolite is present, terminator structure is disrupted, resulting in gene activation. (**b**) RNA-based metabolite biosensors regulate translation. The presence of metabolite activates RBS, leading to gene expression. (**c**) A ribozyme-based metabolite biosensor regulates RNA stability by modulating mRNA cleavage, (**d**) a metabolite biosensor based on RNA splicing. Binding of the metabolite inhibits the splicing, leading to increased gene expression

translational regulation in response to exogenously applied ligand theophylline [69].

5.2.2.3 Stability-Based RNA Sensors

Regulation of RNA stability provides another mechanism through which gene expression can be controlled by introducing either ribozyme self-cleavage or programmed enzymatic processing by RNases (Fig. 5.2c). For example, the theophylline-responsive aptazymes were constructed in *Saccharomyces cerevisiae* by cloning this aptazyme to the 3′ untranslated region of a fluorescent reporter gene of the aptazyme, leading to decreased mRNA self-cleavage activity and enhanced GFP expression [34]. Another strategy is to integrate RNA aptamers with RNase to tune gene expression through directed cleavage of transcripts by an RNase III enzyme. For instance, a class of RNA sensing-actuating devices based on direct integration of an RNA aptamer into a region of the Rnt1p hairpin was constructed to modulate Rnt1p cleavage rates. When theophylline was present, the aptamer bond with theophylline resulted in structural change that inhibits Rnt1p cleavage activity, thus increasing the stability of the transcript [3].

5.2.2.4 Splicing Riboswitch-Based RNA Sensors

In eukaryotic cells, “self-splicing” is typically required to cut out the noncoding introns after transcription. The programmed removals of introns coupled with aptamers within key intronic locations that regulate splicing in response to small molecule provide a critical regulatory approach in the expression of many genes (Fig. 5.2d) [10, 30, 35]. For example, a tetracycline sensor was created by incorporating a tetracycline aptamer in the 5′ splice site in such a way that adding tetracycline facilitates the formation of an aptamer-tetracycline complex structure that inhibits splicing [72].

5.2.3 Protein Activity-Based Sensors

Protein activity-based sensors act independently of translational regulation by directly linking the activity of a screenable or selectable reporter to the binding of a small molecule. Nature contains many examples of sensing by allosteric regulation of protein activity. To be useful for metabolic engineering applications, sensors must bind a ligand relevant to the engineered pathway and transmit this event to a change in the activity of a protein useful for reporting, screening, or regulating other pathway components [35].

5.2.3.1 Combined Domain Sensors

Sensors with desired input and output functions can be generated by combining two independent proteins or protein domains, in such a way that binding of the small molecule ligand to the input component induces a conformational change that alters the enzymatic activity of the output component (Fig. 5.3a). For example, maltose sensors have been reported for increasing of β-lactamase (BLA) activity [22–24]. The maltose-binding domain is selected from a maltose-binding protein (MBP), one of many bacterial periplasmic binding proteins (bPBPs) that bind nutrients, including sugars, ions, and peptides. bPBPs have two domains in a hinge region, where ligand binding to the surface between these domains directs a hinge twist conformational change in the protein. The ligand-binding activity of MBP and selectable activity of BLA were combined into chimeric MBP-BLA proteins by randomly or specifically inserting BLA into MBP. In these sensors, maltose binding to MBP induced a conformational change in the active sensors that allosterically regulated β-lactamase activity and led to increasing cell survival on β-cyclodextrin, thus reporting on the level of maltose in *E. coli*. This sensing system was then further explored to allow detection of new molecules, such as sucrose by mutating the ligand-binding pocket [16, 22].

5.2.3.2 Intein-Based Protein Sensors

A second type of protein activity-based sensor uses inteins, which are segments of proteins that are able to excise themselves and splice the remains. By inserting a ligand-binding domain within the N- and C- termini regions of an intein, small molecule-dependent intein splicing systems could be developed. Then, the sequence is inserted in front of a reporter protein. Binding of the small molecule can either promote or inhibit splicing by influencing the ability of the two intein domains to come together in a conformation that stimulates splicing (Fig. 5.3b). As such, the level of active spliced protein can be used as a readout of small molecule ligand concentration [35].

In one example, by inserting a human estrogen receptor binding domain (ER) between the N- and C- termini, hormone analogue-dependent splicing was engineered into the RecA intein from *Mycobacterium tuberculosis* [9, 55]. In another example, an intein-based biosensor has been constructed based on rational design whose splicing activity is triggered in vivo in response to thyroid hormone or synthetic analogues [56]. Although the only examples of engineered ligand-responsive inteins developed thus far are hormone-responsive and incorporate receptor binding domains, it is plausible that this same design principle could be used to incorporate binding domains for metabolites. Furthermore, once a metabolite-binding intein is developed, it could potentially be inserted into any polypeptide to control processing to an active protein in response to the small molecule ligand. Thus, inteins can be used as small molecule sensors that act post-translationally to control the expression of pathway enzymes within a host cell.

5.2.3.3 Yeast Three-Hybrid Sensors

The yeast three-hybrid system can also be employed as sensing strategy. The traditional yeast three-hybrid system is an extension of the two-hybrid assay to include small molecule-dependent protein-protein interactions. In the yeast three-hybrid system, the two domains of the Gal4 transcription factor, DNA-binding domain and an activating domain, are fused to a

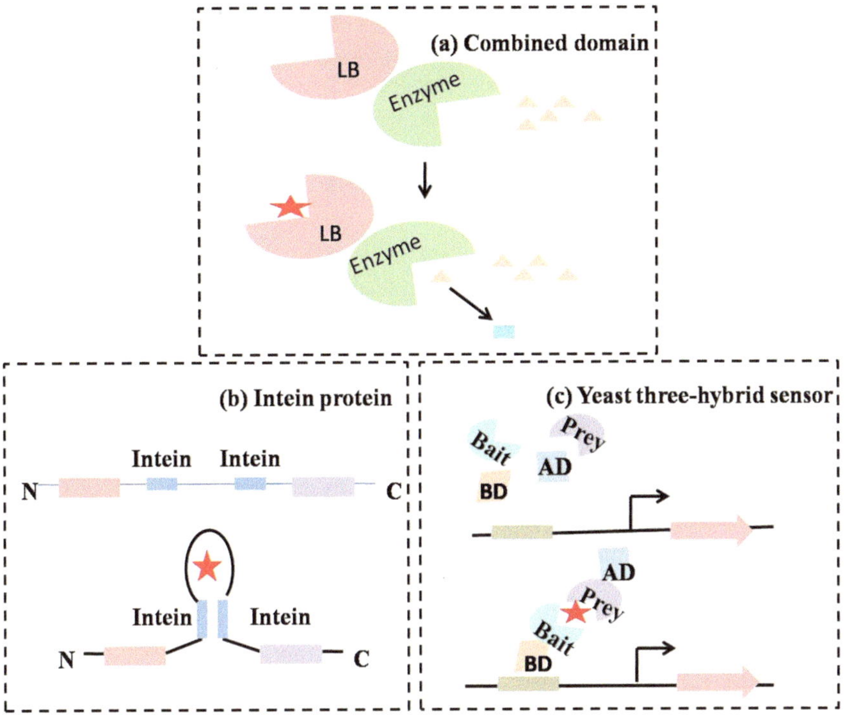

Fig. 5.3 Protein activity-based metabolite biosensors. (**a**) A combined domain-based biosensor regulates protein activity by conformational change of ligand binding (LB) at the presence of metabolite. (**b**) A intein protein-based biosensor uses ligand-dependent intein splicing to link metabolite to regulate protein activity. (**c**) A yeast three-hybrid biosensor regulates gene expression by modulating interactions among prey, bait, and metabolite

bait protein and a library of prey proteins, respectively, such that in the presence of a given small molecule, the protein-protein interaction between bait and prey reconstitutes the transcriptional activator and drives expression of a reporter gene (Fig. 5.3c). This system is readily extended to measure levels of a metabolite by replacing the bait and prey with two known proteins whose binding depends on the target small molecules [35].

One of the three-hybrid sensor designs was tested using retinoid X receptor (RXR) for detection of retinoic acid and its synthetic analogues. To create a new ligand for the receptors, a structure-based approach was used to generate a library of ~380,000 mutant RXR genes. Positive variants were transcriptionally active with improved 25-fold sensitivity comparing to one that was engineered through site-directed mutagenesis [53]. However, the major limitation of the yeast three-hybrid sensor design is that it can only be applied to the detection of a small molecule for which bait-prey protein partners are available, primarily hormone receptors and other cell signaling components. Therefore, despite its sensing capabilities, this class of sensors will have limited utility in metabolically engineered systems except in rare instances [35].

5.2.4 Whole Cell Sensors

In addition to protein- and RNA-based sensors, whole cell sensors based on microbial auxotrophy have been used to report the concentration of growth-limiting small molecules. For example, an engineered *E. coli* mevalonate auxotroph was generated by reporting on the mevalonate concentration in the growth medium through a change in growth rate [35, 42]. By knocking out the native mevalonate pathway and introducing a heterologous operon for the utilization of mevalonate together with an independent GFP reporter gene, the growth rate dependence on mevalonate concentration has been modeled based on the fluorescent readout. Furthermore, a recently reported computational design strategy for generating auxotrophic *E. coli* mutants may expand the number of available cell sensors to as many as 53 small molecules [63]. In general, this method should be applicable to the quantification of other metabolites for which viable auxotrophs can be developed. As one of the goals of metabolic engineering is to produce new molecules from substrates supplied directly by the host cell metabolism, whole cell sensors will be valuable tools for optimizing this concentration between the native primary metabolism of the host cell and the introduced heterologous pathways [35].

5.3 Applications in Metabolic Engineering

5.3.1 Acyl-CoA Precursor to Fatty Acid Ethyl Ester (FAEE) Production

Acyl-CoA is a key intermediate involved in the fatty acid ethyl ester (FAEE) biosynthetic pathway, which is a temporary compound formed by attaching coenzyme A to the end of a long-chain fatty acid inside living cells which then reacts with ethanol to form a FAEE. Since acyl-CoA is a low abundance metabolite, it is difficult to measure accurately by conventional methods. In order to direct more flux to FAEE, building an acyl-CoA-targeted biosensor is necessary. For example, the naturally occurring fatty acid-sensing protein FadR was engineered to upregulate acyl-CoA biosynthesis, ethanol production, and the expression of a wax ester synthase, which direct more flux to form FAEE (Fig. 5.4). As a result, it allows the downstream pathway to be activated only when there is sufficient acyl-CoA and avoids the production of unnecessary proteins and ethanol at the early stage of fermentation. The final FAEE titer was increased to 1.5 g/L and the yield increased threefold to 28 % of the theoretical maximum [77].

5.3.2 Malonyl-CoA Precursor to Fatty Acid Production

Similar to acyl-CoA, malonyl-CoA is a key intermediate in fatty acid biosynthesis and polyketide biosynthesis. It is synthesized from acetyl-CoA by acetyl-CoA carboxylase (encoded by *acc*). Overexpression of *acc* not only improves fatty acid production, but it also inhibits cell growth. To alleviate the inhibitory effect of *acc* overexpression while maintaining high malonyl-CoA concentrations, malonyl-CoA sensors were studied to dynamically downregulate *acc* expression when cells accumulate high malonyl-CoA levels. For example, the malonyl-CoA sensor-actuator has been constructed based on a naturally occurring malonyl-CoA transcription factor, FapR, from the Gram-positive bacteria *Bacillus subtilis*. FapR specifically binds to a 17-bp DNA sequence and negatively regulates fatty acid and phospholipid metabolism in *B. subtilis*. The binding of malonyl-CoA to FapR triggers a conformation change to the FapR, causing FapR-DNA complex to dissociate [31]. Malonyl-CoA source pathway was under the control of malonyl-CoA-downregulated pGAP promoter and malonyl-CoA sink pathway was under the control of malonyl-CoA-upregulated T7 promoter (Fig. 5.5) [73].

5.3.3 Mevalonate Precursor to Terpene and Steroid Production

The mevalonate-dependent isoprenoid pathway converts acetyl coenzyme A (acetyl-CoA) into the five-carbon-atom isoprenoid building block, isopentenyl pyrophosphate (IPP). The reduction of hydroxymethylglutaryl-CoA (HMG-CoA) to mevalonate by HMG-CoA reductase is a key step in this pathway. The MEV pathway is native to eukaryotes and prokaryotes, but not native to *E. coli*. The heterologous MEV operon is composed of *atoB* encoding *E. coli* acetoacetyl-CoA thiolase, ERG13 encoding *Saccharomyces cerevisiae* 3-hydrroxy-3-methylglutaryl-CoA synthase, and a truncated HMG1 gene from *Saccharomyces*

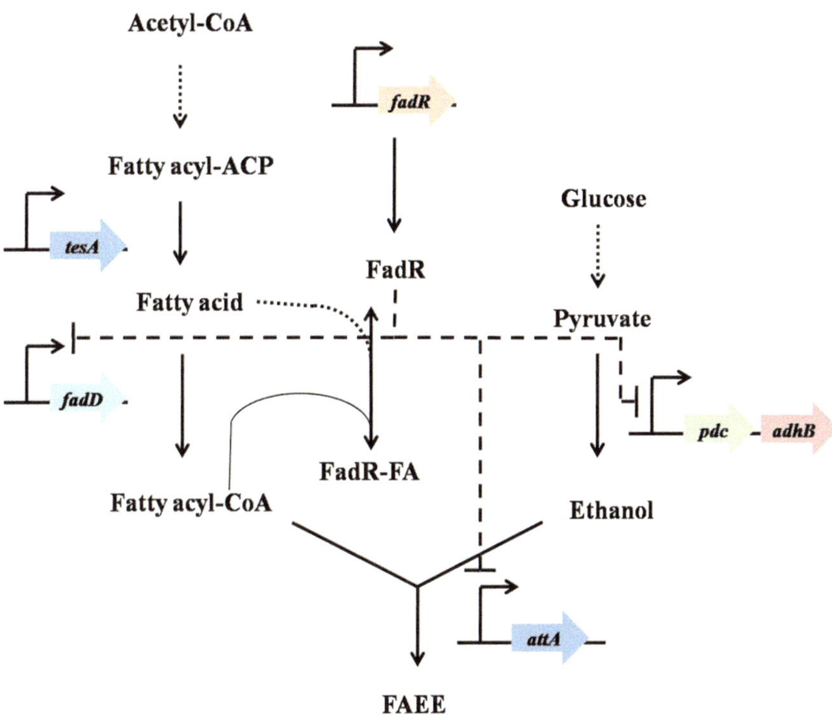

Fig. 5.4 Acyl-CoA biosensors were used for FAEE production by dynamically regulating its downstream enzyme expression

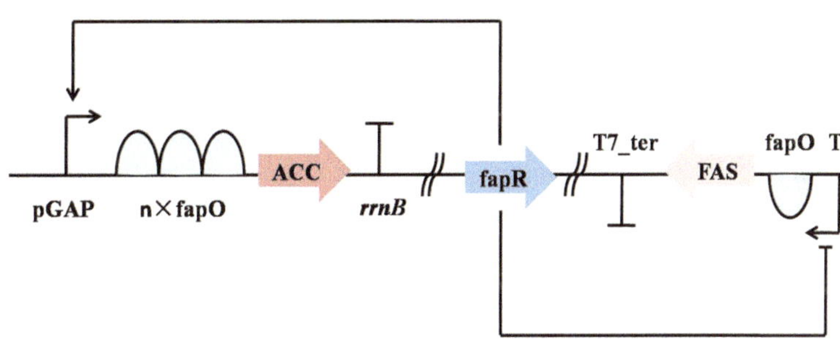

Fig. 5.5 Metabolite biosensor to regulate metabolic pathway for fatty acid production

cerevisiae encoding a soluble version of HMG-CoA reductase. The production of isoprenoids in *E. coli* through the heterologous MEV pathway is limited by mevalonate supply. A mevalonate-responsive AraC variant was isolated and then used to high-throughput screen for improved mevalonate production as a result of MevT-operon mutations [60]. When mevalonate accumulated at 30 mM or higher, mevalonate-AraC complex activated the P_{BAD} promoter and GFP was detected as reporter.

5.3.4 Amino Acid Production

Amino acids are major industrial products derived from fermentation of microorganisms, comprising a world market of more than 3 million tons per year [5]. The Gram-positive bacterium *Corynebacterium glutamicum* alone is used for the industrial production of L-lysine on a scale of 1.3×10^6 tons/year [67]. For example, a FACS high-throughput method has been built to clone *eyfp* (enhanced yellow fluorescent protein) at 3′ of a *Corynebacterium glutamicum* promoter that is regulated by an endogenous transcription factor Lrp, which can detect L-methionine and several branched-chain amino acids, including L-valine, L-leucine, and L-isoleucine [7]. Using chemical mutagens, random mutations were introduced to the *C. glutamicum* strains, which carry the sensor plasmid. Cells cultivated and screened by FACS and the ones with enhanced fluorescence were isolated and recultivated to enrich the high-producing strains. Mutants that produce up to a total of 11 mM branched-chain amino acids were identified using this method.

5.3.5 Triacetic Acid Lactone Production

Triacetic acid lactone (TAL), also referred to as 4-hydroxy-6-methyl-2-pyrone, is a natural compound of polyketide origin, commonly identified as a triketide derailment product during polyketide biosynthesis (e.g., lovastatin and 6-methylsalicylic acid) [32, 48]. TAL is also a precursor in the chemical synthesis of phloroglucinol, used in the synthesis of the thermostable energetic material 1,3,5-triamino-2,4,6-trinitrobenzene (TATB), and resorcinol, used in resin and adhesive formulations [1, 25]. Microbial synthesis of TAL starts from glucose as substrate and 2-pyrone synthase (2-PS), encoded by the *g2ps1* gene isolated from native TAL producer, *Gerbera hybrid* [15]. To date, improving TAL production has been limited by the lack of sensitive and rapid screening/selection methods for identifying desirable candidates from gene libraries [62]. Cirino and coauthors developed a mutated "AraC," which responds to TAL to actuate expression of green fluorescent (gfpuv) from promoter P_{BAD} in *E. coli*. After multiple site saturation mutagenesis of five amino acid located in the AraC binding pocket (P8V, T24I, H80G, Y82L, and H93R), the AraC mutant responded to the presence of exogenous 5 or 2.5 mM TAL, which a high-throughput FACS method was constructed. After two randomly mutated *g2ps1* gene using error-prone PCR, a variant showed around 20-fold increase.

5.3.6 Flavonoid Compounds Production

Naringenin, a pharmacologically useful plant flavonoid molecule was able to be produced from *E. coli* by heterologous expression of four enzymes: tyrosine ammonia lyase, 4-coumaroyl ligase, chalcone synthase, and chalcone isomerase [51]. In a recent paper, Raman et al. used metabolite-responsive transcription factor-regulated promoters to control the expression of TolC, a protein that allows both positive and negative selections when supplemented with sodium dodecyl sulfate (SDS) and colicin E1, respectively. While positive selection was needed to select for high-producing strains generated by multiplex automated genome engineering (MAGE), negative selection was used to eliminate the false positives caused by mutations. This transcription factor-based method was successfully implemented to enhance production for naringenin [45].

5.3.7 Biofuel Production

With depletion of the nonrenewable fossil fuels and increase demand for oil use, microbial production of biofuels has advantages of low cost, high energy, and renewability. However, one challenge is that high biofuel production usually requires host cells to exhibit high tolerance to biofuels [19, 36]. Even though host cells with high tolerance ability can be obtained, it does not necessarily mean that the strain has native high production capability. On the other hand, direct high-throughput methods to detect high production candidates are uncommon. One way to address this problem is to develop biosensors for direct detection of small molecules to rapidly and specifically screen for the desired phenotype. One example has been studied by using a biosensor based on a 1-butanol-responsive transcription factor-promoter pair controlling expression of a tetracycline resistance reporter protein. A putative δ-54-transcriptional activator (BmoR) and a δ-54-dependent, alcohol-regulated promoter (P_{BMO}) were identified in *Pseudomonas butanovora* [13].

5.3.8 Environmental Toxin Detection

Metabolite biosensors can be utilized for environmental toxin detection due to the response of metabolite to specific transcription factor and metabolite-transcription factor complex activate certain reporter production at presence of metabolite. For example, water-soluble aromatic components (e.g., benzene, toluene, ethylbenzene, and xylene) of petroleum products can adversely impact groundwater, depending on local biogeochemical conditions [47]. But they often persist in the environment and are hard to detect. A whole cell bacterial biosensor based on *E. coli* BL21DE3(RIL) expressing gfp under the control of an alcohol dehydrogenase inducible promoter belonging to the archaeon *Sulfolobus solfataricus* (Sso2536adh promoter) was used to measure aqueous concentrations of aromatic aldehydes [18]. The *E. coli* BL21DE3(RIL) biosensor strain displayed a specific response and high sensitivity to the different aromatic aldehydes used, such as benzaldehyde, cinnamaldehyde, and salicylaldehyde, suggesting its potential low-cost application to environmentally relevant samples.

Hydrocarbon pollution represents a widespread problem to native organisms in a wide range of environments, and detection may be possible using alkane-responsive biosensors. An alkane-responsive biosensor with a fluorescence output signal in *Escherichia coli* by using regulatory machinery from alkane metabolism in *Pseudomonas putida* has been developed [46]. Within that system, the transcriptional regulator, AlkSp, is activated by the presence of alkanes and binds to the P_{alkB} promoter, stimulating transcription of a GFP reporter. After two rounds of directed evolution via error-prone PCR and high-throughput screening, an alkS mutant enabled up to a fivefold increase in fluorescence output signal in response to short-chain alkanes such as hexane and pentane.

5.4 Methodologies

5.4.1 Design of Transcription Factor

5.4.1.1 Modification of Natural Transcription Factor

Transcription factors are essential for the regulation of gene expression and are, as a consequence, found in almost all living organisms. Some of the most commonly used transcription factors from nature are AraC, LacI, and FapR, which are regulated by small chemicals such as arabinose, IPTG, and malonyl-CoA, respectively. These naturally occurring transcription factors can be modified or synthetically implemented as regulatory elements to implement toggle switches or oscillators [8, 57]. Engineering transcription factor proteins that control transcription in response to nonnative small molecule stimuli can be used as genetic switches in biosensing and metabolic engineering. For example, Schleif and coworkers have characterized AraC and the mechanisms of the ara operon regulation and proposed the "light switch" mechanism [52]. In the absence of L-arabinose, the DNA-binding domains (DBDs) of an AraC dimer bind the I1 and O2 half-sites (separated by 210 bases), repressing transcription through the formation of a DNA loop upstream transcription through the formation of a DNA loop upstream of the P_{BAD} promoter. Upon binding L-arabinose, the dimer changes conformation such that the DBDs bind the adjacent I1 and I2 half-sites, resulting in transcriptional activation via interactions with RNA polymerase at P_{BAD}. Induction of the ara operon is specific to L-arabinose: structurally and chemically similar sugars such as D-xylose, D-arabinose, and D-fucose (6-deoxy-D-galactose) fail to act as wild-type AraC effectors. Studies found that two sites showed critical interactions including N-terminal AraC arm and the C-terminal DBD in the absence of inducer and the arms and ligand-binding pocket in the presence of L-arabinose.

Mutation of the N-terminal AraC results in constitutive, noninducible expression. Cirino and coworkers successfully modified natural AraC specificity by subjecting five residues of its ligand-binding pocket with saturation mutagenesis [60, 61].

5.4.1.2 De Novo Artificial Transcription Factor (ATF)

Synthetic transcriptional regulators typically bear two essential yet separate modules: the DNA-binding domain (DBD) and the regulatory domain (RD). The DBD imparts most of the specificity in targeting the RD to a particular site in the genome. The RD usually plays a less critical role in selecting a gene for regulation; on the contrary, they mediate their effects directly on the gene to which they are delivered. In order to achieve activation or active repression, synthetic DBD and RD can be linked together to function as an artificial transcription function (ATF) [2, 44]. Two examples of ligand-dependent ATFs were provided by the groups of Bujard and Schreiber. Bujard and coworkers developed ATFs that bind to DNA only in the presence of doxycyclin, whereas Schreiber and coworkers used chemical inducers of dimerization (CIDs) to mediate the interaction of a DBD and a RD in eukaryotes [4, 6]. Another example is that Cornish and coworkers reported a CID composed of methotrexate and a synthetic analogue of the natural product FK506 to manipulate the interaction of a DBD with an activating region that functions robustly in bacteria [6].

One ATF application in metabolic engineering was developed for isoprenoids production by replacing AraC's ligand-binding domain (LBD) with isopentenyl diphosphate isomerase (Idi) that naturally binds isoprenoids. The choice of Idi is reasonable due to crystallographic data indicating that dimerization of Idi could create at least two different conformational states to activate transcription. This approach is useful to develop sensors for tyrosine and isoprenoid production [11].

5.4.2 In Silico Design of Ribozymes

In this section we describe computational methods for designing allosteric ribozymes, especially hammerhead ribozymes (HHRs) that can sense small molecules. In nature, HHRs consist of ligand(s)-binding allosteric domain and a catalytic center. The allosteric ribozymes can be switched on or turned off as a result of binding small molecule or oligonucleotides to the ligand-binding domain. Studies on HHRs have shown that there are three types of methods for obtaining the allosteric ribozymes: (1) in vitro selection [17, 26], (2) rational design [29, 59], and (3) computational selection [40, 41].

The important advantage of using computational design of small molecule-sensing ribozyme over in vitro selection and rational design methods is the possibility to compute all possible random sequences that fuse the aptamer domain to the ribozyme. It provides the possibility to obtain the sequences with the best possible properties for a given length of the communication module. If we are not satisfied with the properties of the obtained sequences, we can easily change the length of the communication module, which is another advantage of using computational design methods. Naturally, the main disadvantage is that computational methods need to be evaluated and tested using experiments.

5.4.2.1 Algorithm-Based Design

There are two approaches for computational design of small molecule-sensing ribozymes. The first approach is to compute the sequence of the communication module between the ribozyme and the aptamer based on the partition function for RNA folding by applying a random search algorithm [40]. For example, a new ribozyme can be generated using a sequence that contains the extended hammerhead motif from *Schistosomes* and the theophylline aptamer. One example of this approach was implemented to design a high-speed allosteric ribozyme with NOT logic function that senses the presence of theophylline, shown in Fig. 5.6 [40].

5.4.2.2 3D Modeling Tertiary Structure-Based Design

A second approach for computational design of small molecule-sensing allosteric ribozymes is based on modeling 3D interactions between the ligand and its RNA aptamer. To apply this approach, tertiary structure of the RNA aptamer bound to the ligand and interactions between them are required. Available tertiary structures can be found in protein structure databases such as the protein data bank (pdb, http://www.pdb.org/). For example, purine-sensing ribozymes were designed by inserting guanine and adenine aptamers into the minimal version of the HHR based on 3D structures of corresponding purine riboswitches found in bacteria. Molecular dynamic simulations were then carried out by using Amber suite (http://amber.script.edu/) to calculate interactions between the guanine and aptamers that are embedded into the stem II of the ribozymes. Opposite logic functions (YES or NOT) were constructed at only one base pair difference, and both were experimentally tested in the presence or absence of guanine [39].

5.4.3 Design of Protein Sensor

Three branches of protein engineering can be identified: rational design that uses site-directed mutagenesis to modify existing proteins, de novo design that involves the synthesis of new protein from first principles by using established knowledge on protein folding and structure prediction, and directed evolution that uses random mutagenesis on known gene sequences to generate new proteins or enzymes to achieve a new target function in a fortuitous manner identified by screening or selection of a wide range of sequences [20].

5.4.3.1 Rational Design

Rational protein engineering is based on site-directed mutagenesis and relies on existing information of the 3D structure of the target protein and the implication of specific residues in its function. Protein modeling is a discipline in its own right, and it relies on the vast amount of structural and functional information stored in databases available. Once the 3D X-ray or NMR

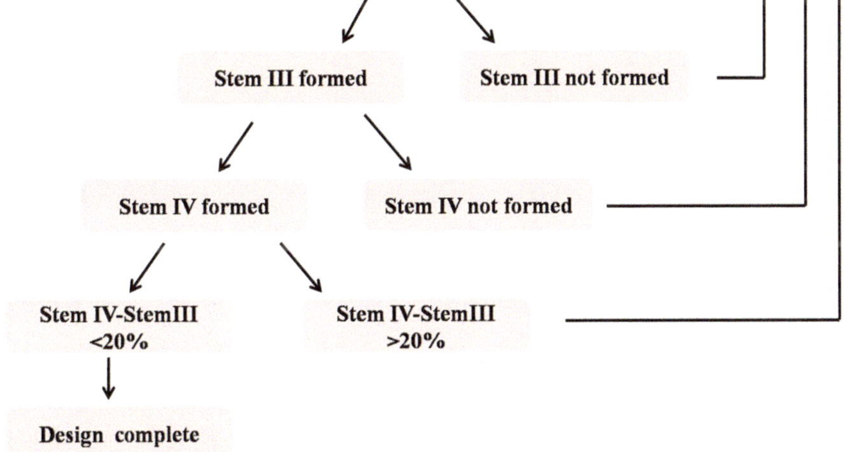

Fig. 5.6 Computational selection hammerhead ribozymes that sense small molecules

structure for the target protein has been obtained, molecular modeling is used to locate the key residues to be targeted by mutagenesis, as well as to perform calculations to interpret or extrapolate principles for design of the mutants. Site-directed mutagenesis can be used to replace, delete, or insert one or more amino acids by introducing mutagenic primers to accumulate exponentially at each cycle [50].

5.4.3.2 De Novo Synthesis of Protein Biosensors

Proteins can also be created from first principles by solid state synthesis [20, 38]. Based on understanding hydrophobic effects driving folding and intra-chain hydrogen binding patterns, computational design of active proteins have been achieved for a variety of reactions including the Diels-Alder reaction, the Kemp elimination, and the retro-aldol reaction. For example, Rosetta de novo enzyme design has been used to design enzyme catalysts for different chemical reactions. It includes four stages: (1) choice of a catalytic mechanism and corresponding minimal model active site, (2) identification of sites in a set of scaffold proteins where this minimal active site can be realized, (3) optimization of the identities of the surrounding residues for stabilizing interactions with the transition state and primary catalytic residues, and (4) evaluation and ranking the resulting designed sequences [49].

5.4.3.3 Protein Design by Directed Evolution

Directed evolution is based on a number of cycles of random mutagenesis aiming at achieving new functions in existing proteins such as high chemical and thermal stability, solubility in organic solvents, activity toward new substrates, and enantio- or regioselectivity in catalysis. Basically, directed in vitro evolution mimics the process of natural molecular evolution with four main steps: choosing a parent protein, creating a mutant library based on the parent protein, identifying variants with improved target properties, and repeating the entire process until achieving the desired function, also referred as SELEX [71, 75]. Error-prone PCR was introduced to produce the mutagenesis libraries by using DNA polymerases lacking proofreading activity, such as Taq polymerase from *Thermus aquaticus*, Vent polymerase from *Thermococcus litoralis*, and Pfu from *Pyrococcus furiosus*. The number of possible variants (V) of a protein that can be created by introducing M substitutions simultaneously over N amino acids could be estimated using the equation below [20]:

$$V = \frac{N!19^M}{(N-M)!M!}$$

It could be estimated that there are 177,848 possible variants with only 9 targeted positions over 4 amino acid changes simultaneously.

Another key component is a fast, sensitive, and specific high-throughput screening assay to enable the identification of positive variants. The development of the screening method is usually the critical bottleneck step in the directed evolution of a particular enzyme. Although X-ray crystallography and NMR spectroscopy offer a detailed analysis of the variant based on the structure and function, the applications of these techniques are not always quick and straightforward. Alternatively, circular dichroism, fluorescence spectroscopy, and calorimetry methods provide useful information to quickly detect the active site of enzymes [71].

5.5 Future Perspectives

Recent research has contributed major innovations in the development of metabolite biosensors with increasing numbers of metabolite targets, mechanisms of action, and applications in metabolic engineering. However, in order to maximize the potential of this emerging technology, many challenges must be addressed. One consideration involves the chemical nature of the metabolite-binding domain. For example, the limited diversity of available RNA parts is a major constraint in the application of nucleic acid-based sensors, although design of ribozyme technologies (intro selection, rational design, and computational design) may allow rapid exploration of the functional sequence space. On the other hand, linking metabolite binding to novel, desirable changes in protein properties is substantially more challenging. Protein folding, metabolite-binding-induced conformational changes, and intra- or intermolecular signal transduction are currently harder to predict and engineer than nucleic acid-based chemistry. Successful approaches often require multiple rounds of complementary computational, experimental, and directed evolution approaches. This may be one reason why metabolite biosensors have not expanded into some applications that may be useful for dynamic regulation. Second, introducing synthetic RNAs and proteins may potentially cause an increase in cellular "burden" as cellular resources are shared between production synthesis and cellular growth. From this perspective, RNA-based metabolite biosensors tend to be superior to protein activity-based and transcription factor-based biosensor due to a lack of translation and posttranslation modification of target protein. A third area of concern is the temporal delay associated with the response time from metabolite sensing to actuation, because biosensors inherently have a time lag between the true metabolite level changes and the downstream effects associated with regulating transcription or translation levels . For example, protein activity-based sensors respond to metabolite level changes faster than RNA-based sensors or transcription-based sensors. Thus, protein activity-based biosensors may be a good fit in sensing those relatively toxic, high-flux metabolic intermediates or selecting high-producing candidates by high-throughput method. However, for those relatively stable and slow-changing metabolites, drastic changes in metabolite levels may not be desirable.

References

1. Agrawai J (1998) Recent trends in high energy materials. Prog Energ Combust 24(1):1–30
2. Ansari AZ, Mapp AK (2002) Modular design of artificial transcription factors. Curr Opin Chem Biol 6:765–772
3. Babiskin AH, Smolke CD (2011) Engineering ligand-responsive RNA controllers in yeast through the assembly of RNase III tuning modules. Nucleic Acids Res 39(12):5299–5311. doi:10.1093/nar/gkr090
4. Baron U, Bujard H (2000) Tet repressor-based system for regulated gene expression in eukaryotic cells: principles and advances. Methods Enzym 327:401–421
5. Becker J, Zelder O, Hafner S, Schroder H, Wittmann C (2011) From zero to hero – design-based systems metabolic engineering of *Corynebacterium glutamicum* for L-lysine production. Metab Eng 13(2):159–168. doi:10.1016/j.ymben.2011.01.003
6. Belshaw PJ, Ho SN, Crabtree GR, Schereiber SL (1996) Controlling protein association and subcellular localization with a synthetic ligand that induces heterodimerization of proteins. Proc Natl Acad Sci U S A 93(10):4604–4607
7. Binder S, Schendzielorz G, Stabler N, Krumbach K, Hoffmann K, Bott M, Eggeling L (2012) A high-throughput approach to identify genomic variants of bacterial metabolite producers at the single-cell level. Genome Biol 13(5):R40. doi:10.1186/gb-2012-13-5-r40
8. Broun P (2004) Transcription factors as tools for metabolic engineering in plants. Curr Opin Plant Biol 7(2):202–209. doi:10.1016/j.pbi.2004.01.013
9. Buskirk AR, Ong YC, Gartner ZJ, Liu DR (2004) Directed evolution of ligand dependence: small-molecule-activated protein splicing. Proc Natl Acad Sci U S A 101(29):10505–10510. doi:10.1073/pnas.0402762101
10. Chang AL, Wolf JJ, Smolke CD (2012) Synthetic RNA switches as a tool for temporal and spatial control over gene expression. Curr Opin Biotechnol 23(5):679–688. doi:10.1016/j.copbio.2012.01.005
11. Chou HH, Keasling JD (2013) Programming adaptive control to evolve increased metabolite production. Nat Commun 4:2595. doi:10.1038/ncomms3595
12. Dahl RH, Zhang F, Alonso-Gutierrez J, Baidoo E, Batth TS, Redding-Johanson AM, Petzold CJ, Mukhopadhyay A, Lee TS, Adams PD, Keasling JD (2013) Engineering dynamic pathway regulation using stress-response promoters. Nat Biotechnol 31(11):1039–1046. doi:10.1038/nbt.2689
13. Dietrich JA, Shis DL, Alikhani A, Keasling JD (2013) Transcription factor-based screens and synthetic selections for microbial small-molecule biosynthesis. ACS Synth Biol 2(1):47–58. doi:10.1021/sb300091d
14. Dixon N, Duncan JN, Geerlings T, Dunstan MS, McCarthy JE, Leys D, Micklefield J (2010) Reengineering orthogonally selective riboswitches. Proc Natl Acad Sci U S A 107(7):2830–2835. doi:10.1073/pnas.0911209107
15. Eckermann S, Schröder G, Schmidt J, Strack D, Edrada RA, Helaiutta Y, Elomaa P, Kotilainen M, Kilpeläinen I, Proksch P, Teeri TH, Schröder J (1998) New pathway to polyketides in plants. Nature 396:387–390. doi:10.1038/24652
16. Edwards WR, Busse K, Allemann RK, Jones DD (2008) Linking the functions of unrelated proteins using a novel directed evolution domain insertion

method. Nucleic Acids Res 36(13), e78. doi:10.1093/nar/gkn363
17. Ellington AD, Szostak JW (1990) In vitro selection of RNA molecules that bind specific ligands. Nature 340:818–822
18. Fiorentino G, Ronca R, Bartolucci S (2009) A novel E. coli biosensor for detecting aromatic aldehydes based on a responsive inducible archaeal promoter fused to the green fluorescent protein. Appl Microbiol Biotechnol 82(1):67–77. doi:10.1007/s00253-008-1771-0
19. Fischer CR, Klein-Marcuschamer D, Stephanopoulos G (2008) Selection and optimization of microbial hosts for biofuels production. Metab Eng 10(6):295–304. doi:10.1016/j.ymben.2008.06.009
20. Gilardi G (2013) Protein design for biosensors. In: Roberts GCK (ed) Encyclopedia of biophysics. Springer, Berlin, pp 1979–1987. doi:10.1007/978-3-642-16712-6
21. Gilbert W, Müller-Hill B (1966) Isolation of the Lac repressor. Proc Natl Acad Sci U S A 56(6):1891–1898
22. Guntas G, Mansell TJ, Kim JR, Ostermeier M (2005) Directed evolution of protein switches and their application to the creation of ligand-binding proteins. Proc Natl Acad Sci U S A 102(32):11224–11229. doi:10.1073/pnas.0502673102
23. Guntas G, Mitchell SF, Ostermeier M (2004) A molecular switch created by in vitro recombination of nonhomologous genes. Chem Biol 11(11):1483–1487. doi:10.1016/j.chembiol.2004.08.020
24. Guntas G, Ostermeier M (2004) Creation of an allosteric enzyme by domain insertion. J Mol Biol 336(1):263–273. doi:10.1016/j.jmb.2003.12.016
25. Hansen CA, Frost J (2002) Deoxygenation of polyhydroxybenzenes: an alternative strategy for the benzene-free synthesis of aromatic chemicals. J Am Chem Soc 124(21):5926–5927
26. Joyce GF (2007) Forty years of in vitro evolution. Angew Chem Int Ed Engl 46(34):6420–6436. doi:10.1002/anie.200701369
27. Keasling JD (2012) Synthetic biology and the development of tools for metabolic engineering. Metab Eng 14(3):189–195. doi:10.1016/j.ymben.2012.01.004
28. Kosuri S, Church GM (2014) Large-scale de novo DNA synthesis: technologies and applications. Nat Methods 11(5):499–507. doi:10.1038/nmeth.2918
29. Liang J, Smolke C (2012) Rational design and tuning of ribozyme-based devices. Methods Mol Biol 848:439–454
30. Liu D, Evans T, Zhang F (2015) Applications and advances of metabolite biosensors for metabolic engineering. Metab Eng 31:35–43. doi:10.1016/j.ymben.2015.06.008
31. Liu D, Xiao Y, Evans BS, Zhang F (2015) Negative feedback regulation of fatty acid production based on a malonyl-CoA sensor-actuator. ACS Synth Biol 4(2):132–140. doi:10.1021/sb400158w
32. Ma SM, Li JW, Choi J, Zhou H, Lee KM, Moorthie VA, Xie X, Kealey JT, Da Silva NA, Vederas JC, Tang Y (2009) Complete reconstitution of a highly reducing iterative polyketide synthase. Science 326:589–592
33. Meinhardt S, Manley MW Jr, Becker NA, Hessman JA, Maher LJ 3rd, Swint-Kruse L (2012) Novel insights from hybrid LacI/GalR proteins: family-wide functional attributes and biologically significant variation in transcription repression. Nucleic Acids Res 40(21):11139–11154. doi:10.1093/nar/gks806
34. Michener JK, Smolke CD (2012) High-throughput enzyme evolution in *Saccharomyces cerevisiae* using a synthetic RNA switch. Metab Eng 14(4):306–316. doi:10.1016/j.ymben.2012.04.004
35. Michener JK, Thodey K, Liang JC, Smolke CD (2012) Applications of genetically-encoded biosensors for the construction and control of biosynthetic pathways. Metab Eng 14(3):212–222. doi:10.1016/j.ymben.2011.09.004
36. Mukhopadhyay A (2015) Tolerance engineering in bacteria for the production of advanced biofuels and chemicals. Trends Microbiol 23(8):498–508. doi:10.1016/j.tim.2015.04.008
37. Muranaka N, Sharma V, Nomura Y, Yokobayashi Y (2009) An efficient platform for genetic selection and screening of gene switches in *Escherichia coli*. Nucleic Acids Res 37(5), e39. doi:10.1093/nar/gkp039
38. Nilsson BL, Soellner MB, Raines RT (2005) Chemical synthesis of proteins. Annu Rev Biophys Biomol Struct 34:91–118. doi:10.1146/annurev.biophys.34.040204.144700
39. Penchovsky R (2013) Computational design and biosensor applications of small molecule-sensing allosteric ribozymes. Biomacromolecules 14(4):1240–1249. doi:10.1021/bm400299a
40. Penchovsky R (2014) Computational design of allosteric ribozymes as molecular biosensors. Biotechnol Adv 32(5):1015–1027. doi:10.1016/j.biotechadv.2014.05.005
41. Penchovsky R, Breaker RR (2005) Computational design and experimental validation of oligonucleotide-sensing allosteric ribozymes. Nat Biotechnol 23(11):1424–1433. doi:10.1038/nbt1155
42. Pfleger BF, Pitera DJ, Newman JD, Martin VJ, Keasling JD (2007) Microbial sensors for small molecules: development of a mevalonate biosensor. Metab Eng 9(1):30–38. doi:10.1016/j.ymben.2006.08.002
43. Polizzi KM, Kontoravdi C (2015) Genetically-encoded biosensors for monitoring cellular stress in bioprocessing. Curr Opin Biotechnol 31:50–56. doi:10.1016/j.copbio.2014.07.011
44. Purcell O, Peccoud J, Lu TK (2014) Rule-based design of synthetic transcription factors in eukaryotes. ACS Synth Biol 3(10):737–744. doi:10.1021/sb400134k
45. Raman S, Rogers JK, Taylor ND, Church GM (2014) Evolution-guided optimization of biosynthetic pathways. Proc Natl Acad Sci U S A 111(50):17803–17808. doi:10.1073/pnas.1409523111
46. Reed B, Blazeck J, Alper H (2012) Evolution of an alkane-inducible biosensor for increased responsive-

ness to short-chain alkanes. J Biotechnol 158(3):75–79. doi:10.1016/j.jbiotec.2012.01.028

47. Riccardi C, Di Filippo P, Pomata D, Incoronato F, Di Basilio M, Papini MP, Spicaglia S (2008) Characterization and distribution of petroleum hydrocarbons and heavy metals in groundwater from three Italian tank farms. Sci Total Environ 393(1):50–63. doi:10.1016/j.scitotenv.2007.12.010
48. Richardson M, Pohl N (1999) Tolerance and specificity of recombinant 6-methylsalicyclic acid synthase. Metab Eng 1:180–187
49. Ritcher F, Leaver-Fay A, Khare SD, Bjelic S, Baker D (2011) De novo enzyme design using Rosetta3. PLoS One 6(5):e19230–e19242. doi:10.1371/journal.pone.0019230.g001
50. Sadeghi SJ, Meirinhos R, Catucci G, Dodhia VR, Nardo GD, Gilardi G (2010) Direct electrochemistry of drug metabolizing human flavin-containing monooxygenase: electrochemical turnover of benzydamine and tamoxifen. J Am Chem Soc 132:458–459. doi:10.1021/ja909261p
51. Santos CN, Koffas M, Stephanopoulos G (2011) Optimization of a heterologous pathway for the production of flavonoids from glucose. Metab Eng 13(4):392–400. doi:10.1016/j.ymben.2011.02.002
52. Schleif R, Lis J (1967) The regulatory region of the L-arabinose operon: a physical, genetic and physiological study. J Mol Biol 95:417–431
53. Schwimmer LJ, Rohatgi P, Azizi B, Seley KL, Doyle DF (2004) Creation and discovery of ligand-receptor pairs for transcriptional control with small molecules. Proc Natl Acad Sci U S A 101(41):14707–14712. doi:10.1073/pnas.0400884101
54. Shis DL, Hussain F, Meinhardt S, Swint-Kruse L, Bennett MR (2014) Modular, multi-input transcriptional logic gating with orthogonal LacI/GalR family chimeras. ACS Synth Biol 3(9):645–651. doi:10.1021/sb500262f
55. Skretas G, Wood DW (2005) A bacterial biosensor of endocrine modulators. J Mol Biol 349(3):464–474. doi:10.1016/j.jmb.2005.04.009
56. Skretas G, Wood DW (2005) Regulation of protein activity with small-molecule-controlled inteins. Protein Sci 14(2):523–532. doi:10.1110/ps.04996905
57. Sowa SW, Gelderman G, Contreras LM (2015) Advances in synthetic dynamic circuits design: using novel synthetic parts to engineer new generations of gene oscillations. Curr Opin Biotechnol 36:161–167. doi:10.1016/j.copbio.2015.08.020
58. Stephanopoulos G (2012) Synthetic biology and metabolic engineering. ACS Synth Biol 1(11):514–525. doi:10.1021/sb300094q
59. Tang J, Breaker RR (1997) Rational design of allosteric ribozymes. Chem Biol 4:453–459
60. Tang SY, Cirino PC (2011) Design and application of a mevalonate-responsive regulatory protein. Angew Chem Int Edit 50(5):1084–1086. doi:10.1002/anie.201006083
61. Tang SY, Fazelinia H, Cirino PC (2008) AraC regulatory protein mutants with altered effector specificity. J Am Chem Soc 130:5267–5271
62. Tang SY, Qian S, Akinterinwa O, Frei CS, Gredell JA, Cirino PC (2013) Screening for enhanced triacetic acid lactone production by recombinant *Escherichia coli* expressing a designed triacetic acid lactone reporter. J Am Chem Soc 135(27):10099–10103. doi:10.1021/ja402654z
63. Tepper N, Tomer S (2011) Computational design of auxotrophy-dependent microbial biosensors for combinatorial metabolic engineering experiments. PLoS ONE 6, e16274. doi:10.1371/journal.pone.0016274.g001
64. Topp S, Reynoso CM, Seeliger JC, Goldlust IS, Desai SK, Murat D, Shen A, Puri AW, Komeili A, Bertozzi CR, Scott JR, Gallivan JP (2010) Synthetic riboswitches that induce gene expression in diverse bacterial species. Appl Environ Microbiol 76(23):7881–7884. doi:10.1128/AEM.01537-10
65. Trausch JJ, Ceres P, Reyes FE, Batey RT (2011) The structure of a tetrahydrofolate-sensing riboswitch reveals two ligand binding sites in a single aptamer. Structure 19(10):1413–1423. doi:10.1016/j.str.2011.06.019
66. Turner AP, Karube I, Wilson GS (1986) Biosensors fundamentals and applications. Oxford University, Oxford
67. van Ooyen J, Noack S, Bott M, Reth A, Eggeling L (2012) Improved L-lysine production with *Corynebacterium glutamicum* and systemic insight into citrate synthase flux and activity. Biotechnol Bioeng 109:2070–2081. doi:10.1002/bit.24486/abstract
68. Vandenberg E, Brown R, Krull U (1994) Immobilization of proteins for biosensors development. In: Veliky IA, McLean R (eds) Immobilized biosystems theory and practical applications. Springer, Dordrecht, pp 129–231
69. Verhounig A, Karcher D, Bock R (2010) Inducible gene expression from the plastid genome by a synthetic riboswitch. Proc Natl Acad Sci U S A 107(14):6204–6209. doi:10.1073/pnas.0914423107
70. Wachsmuth M, Findeiss S, Weissheimer N, Stadler PF, Morl M (2013) De novo design of a synthetic riboswitch that regulates transcription termination. Nucleic Acids Res 41(4):2541–2551. doi:10.1093/nar/gks1330
71. Wang M, Si T, Zhao H (2012) Biocatalyst development by directed evolution. Bioresour Technol 115:117–125. doi:10.1016/j.biortech.2012.01.054
72. Weigand JE, Suess B (2007) Tetracycline aptamer-controlled regulation of pre-mRNA splicing in yeast. Nucleic Acids Res 35(12):4179–4185. doi:10.1093/nar/gkm425
73. Xu P, Li L, Zhang F, Stephanopoulos G, Koffas M (2014) Improving fatty acids production by engineering dynamic pathway regulation and metabolic control. Proc Natl Acad Sci U S A 111(31):11299–11304. doi:10.1073/pnas.1406401111
74. Xu P, Wang W, Li L, Bhan N, Zhang F, Koffas MA (2014) Design and kinetic analysis of a hybrid promoter-regulator system for malonyl-CoA sensing in *Escherichia coli*. ACS Chem Biol 9(2):451–458. doi:10.1021/cb400623m

75. Yan Q, Fong SS (2015) Bacterial chitinase: nature and perspectives for sustainable bioproduction. Bioresour Bioprocess 2:31–39. doi:10.1186/s40643-015-0057-5
76. Yuan L, Grotewold E (2015) Metabolic engineering to enhance the value of plants as green factories. Metab Eng 27:83–91. doi:10.1016/j.ymben.2014.11.005
77. Zhang F, Carothers JM, Keasling JD (2012) Design of a dynamic sensor-regulator system for production of chemicals and fuels derived from fatty acids. Nat Biotechnol 30(4):354–359. doi:10.1038/nbt.2149
78. Zhang GC, Liu JJ, Kong II, Kwak S, Jin YS (2015) Combining C6 and C5 sugar metabolism for enhancing microbial bioconversion. Curr Opin Chem Biol 29:49–57. doi:10.1016/j.cbpa.2015.09.008
79. Zhou LB, Zeng AP (2015) Exploring lysine riboswitch for metabolic flux control and improvement of L-lysine synthesis in *Corynebacterium glutamicum*. ACS Synth Biol 4(6):729–734. doi:10.1021/sb500332c

Sustainable Assessment on Using Bacterial Platform to Produce High-Added-Value Products from Berries through Metabolic Engineering

6

Lei Pei and Markus Schmidt

6.1 Introduction

Berries are rich resources of secondary metabolites, particularly known for diverse phenolic compounds. These highly bioactive compounds can be developed into novel nutraceutical and pharmaceutical products, as well as high-added-value natural food additives. Compounds extracted from berries have, e.g., been used as colorants (e.g., anthocyanins) [1]. Meanwhile, some phenolics present in berries are of high added value due to their potential to develop into anticancer drugs (e.g., phenolic acids, flavonols, and flavanols) [2]. The antioxidation properties from berries also make them attractive research subject to develop more efficient nutraceutical products than the current crude extraction formulas (e.g., NutriPhy® Bilberry 100 from Chr. Hansen) [3, 4]. To exploit the full potential of the phenolic molecules from berries, a number of research projects have been conducted ranging from identification of bioactive compounds and elucidation of metabolic pathways (metabolic engineering them into suitable industrial production host cells) to eventually commercial production [3, 5–12].

The cultivation of berries is limited by climate, soil type, and geographic conditions. Just as any other crop, berry plants cannot be cultivated everywhere in the world. Berry production is concentrated in certain regions, which, at the same time, also limits the applications of berry fruits [13]. Providing health benefits of berries to people around the world and year round (off-harvest season), solutions other than direct consumption of berry fruits must be found, such as better dissecting the potent compounds from the berries that are responsible for the claimed health beneficial effects and producing these compounds in a sustainable manner, which fall into the theme of the Sustainable Development Strategy (SDS) of European Union and its Member States drawn upon based on the Agenda 21, a nonbinding, voluntarily implemented action plan of the United Nations on sustainable development [14].

Sustainable development aims to meet the needs of the current generations without harming those of the future generations. It intends to fulfill both immediate and long-term objectives of humanity. The European Union and its Member States have developed the SDS with reviews and revisions constantly [14]. Most of the SDSs have listed the selected indicators, aiming to provide measurements for the degree of sustainability. Even though there is no agreed-upon sustainability assessment framework for bio-based products, there is a common theme that such assessment should be based on assessments on environmental, economic, and social sustainability [15].

L. Pei (✉) • M. Schmidt
Biofaction KG, Vienna, Austria
e-mail: pei@biofaction.com

S. Singh (ed.), *Systems Biology Application in Synthetic Biology*,
DOI 10.1007/978-81-322-2809-7_6

Here we will review the current research developments in exploiting the berry resource to produce high-added-value products for food additives, nutraceuticals, and pharmaceuticals. The existing datasets, methods, and models will be applied to illustrate how to access the sustainability of industrial biocatalytic processes to produce berry phenolics as food additives, nutraceuticals, and pharmaceuticals.

6.2 Current Development on Biocatalytic Processes to Produce High-Added-Value Products from Berries

6.2.1 Berry Genome Databases Have Been Developed to Identify the Novel Berry Phenolics

Although berries have been known for a long time for their applications in food and associated health benefits, there is still a lack of comprehensive study on the full potentials of berries in food, nutraceuticals, and pharmaceuticals. Genetic databases on the berries are required to explore the known and novel potentials of berries.

In the past 20 years, genomics on berry species has been developed, focusing on developing molecular markers to identify berries, using techniques ranging from isoenzymes and restriction fragment length polymorphism (RFLP), arbitrary polymerase chain reaction (PCR)-based markers, and sequence-characterized PCR-based markers to array-based and second-generation sequencing-based single-nucleotide polymorphism (SNP) marker characterization [16]. To better explore the existing molecular genome databases on berries, screening techniques also need to develop. One of such techniques is the SMART high-throughput screening platform. The SMART screening platform has been used to carry out *in vivo* assays performed in yeast cells harboring a specific human disease gene, or its yeast homologue has been developed by using green fluorescent protein constructs controlled by galactose-inducible promoters coupled to fluorescence microscopy, and growth assays allowed the identification of candidate extracts inhibiting pathological processes affecting disease protein subcellular dynamics and cellular growth [17]. Potential bioactivities of the berries can be screened for their pharmaceutical potentials for Parkinson's disease, Huntington's disease, Alzheimer's disease, etc. The yeast two-hybrid approach has been used to screen for compounds interfering with specific protein-protein interactions controlling cell proliferation and cancer processes [18]. This technique can be applied to screen the potential anti-inflammatory properties of berry extracts. Antimicrobial activity of berry extracts have been investigated in common pathogens, such as Gram-negative bacteria (e.g., *E. coli*, *S. poona*, and *P. aeruginosa*) and Gram-positive bacteria (e.g., *S. aureus*, *B. cereus*, *E. faecalis*, and *L. monocytogenes*) [19, 20]. For the antibiotic potentials of berries, the minimal inhibitory concentration (MIC) should be determined on the berry extracts, and the potent compounds responsible for the antibiotic activities should be further identified.

6.2.2 Metabolic Engineering on Industrial Host Cell to Produce Berry Phenolics

Up to date, more than 200 plant genes encoding enzymes for the phenolic biosynthetic pathway, its regulation, and the decoration of its products have been identified [12, 21–23]. Yet identifying specific regulators and decorating enzymes of target berry species and the bioactive compounds of interest remain a challenge. The knowledge on the transcriptome profiles of berry would provide useful insight to study the metabolic pathways [24].

The biosynthesis of resveratrol was engineered and expressed in microbes, such as *Lactococcus lactis* [25]. Resveratrol is one of the bioactive compounds from berry that has been expensively studied due to its cancer chemopreventive actives [26]. A lactococcal model strain with improved intracellular malonyl-CoA expression will be used for production of phenolic compounds [27, 28]. Fisetin is a polyphenol compound with the potential to be developed into anticancer, antiviral, and antiaging drugs and, more recently, known for preventing Alzheimer's disease and type I diabetes. Yet the production of fisetin from extraction is costly and dependent on the unpredictable berry fruit harvest, while the chemical synthesis of fisetin requires the use of toxic chemicals. Therefore, biosynthesis of fisetin via heterogeneous expression in microorganisms would be an eco friendly solution. A metabolic pathway to produce fisetin from L-tyrosine has been expressed in *E. coli* and will be further constructed in *L. lactis* [29]. L-tyrosine is used as a precursor to produce para (p)-coumaric acid. This compound can subsequently be converted into p-coumaroyl-coenzymeA (CoA) by tyrosine ammonialyase and 4-coumaroyl-CoA ligase. In the presence of chalcone synthase and chalcone reductase, one molecule of p-coumaroyl-CoA and three molecules of malonyl-CoA can be converted into isoliquiritigenin. And through three further steps, it can be converted to fisetin.

Other than *L. lactis* and *E. coli*, a prophage-free variant of the wild-type strain *C. glutamicum* ATCC 13032 can act as a chassis strain for pathway metabolic engineering as well. *C. glutamicum* is known for its ability to harness phenylpropanoids [30]. Two clusters of genes are responsible for phenylpropanoid catabolism [30, 31]. Deletion of these catabolic related genes might make the mutant strains unable to degrade phenylpropanoids, which would probably enhance the accumulation of the polyphenolic compounds in the engineered strains.

6.3 Sustainable Assessment Based on Environmental Impacts

The environmental sustainability is an important assessment for a product on its contribution to the sustainable development. Quantitative environmental assessments are critical to assess the actual environmental benefits of the high-added-value phenolic products by biocatalytic processes. Life Cycle Assessment (LCA) has become a more common approach these days, while Environmental Impact Assessment (EIA) is limited to a few cases due to two important drawbacks comparing to LCA: lack of full supply chain analysis and single-factor measurement [32].

LCA is a quantitative tool to assess the sustainability of a product throughout cradle to grave, starting from raw material acquisition, material processing, and product manufacturing, distribution, and use to the end of life (Fig. 6.1a). The general indicators to assess the environmental impacts include resource use, human health, and ecological consequences [33, 34]. LCA is to assess the environmental aspects and potential impacts associated with a product, by compiling an inventory of relevant inputs and outputs of a product, evaluating the potential environmental impacts, and eventually interpreting the results of the inventory analysis. To conduct LCA on high-added-value phenolic products of berry via biocatalytic processes developed by metabolic engineering, the analysis should be conducted as shown in Fig. 6.1c. It will be compared with the conventional approaches (mainly on extraction methods) as shown in Fig. 6.1b.

To provide guidance to conduct LCA on a product, the ILCA handbook was developed by the Institute for Environment and Sustainability in the European Commission Joint Research Centre (JRC) [35]. The Life Cycle Impact Assessment (LCIA) was also developed by JRC to help interpret emissions and resource consumption data that are associated with a product's life cycle in terms of environmental burdens, human health, and resources [36–38] (Fig. 6.2).

The high-added-value compounds of berries usually stored in berry fruits are enclosed in complex insoluble tissues such as vacuoles or lipoprotein bilayers, which can require harsh treatment to release them. The five-stage Universal Recovery Process (URP) is commonly used in the industrial scale of recovery of valuable compounds from berries. It is based on stage of raw material pretreatment, macro- and micromolecule separation, extraction, purification, and product formation [39]. Among these stages, conventional extraction technologies applied to berries to obtain high-added-value phenolics usu-

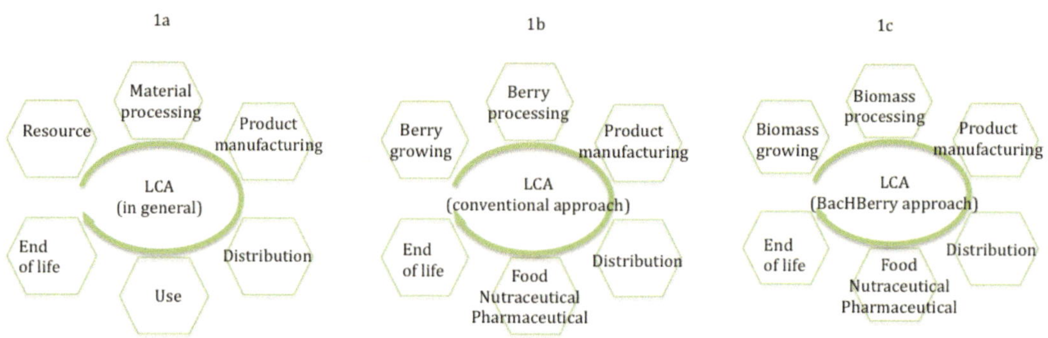

Fig. 6.1 LCA on product in general (**1a**), for using conventional approaches to develop useful products from berries (**1b**), and for approaches developed by BacHBerry an EC-Framework Programme 7 project on "BACterial Hosts for production of Bioactive phenolics from bERRY fruits" (see project website:http://www.bachberry.eu) (**1c**)

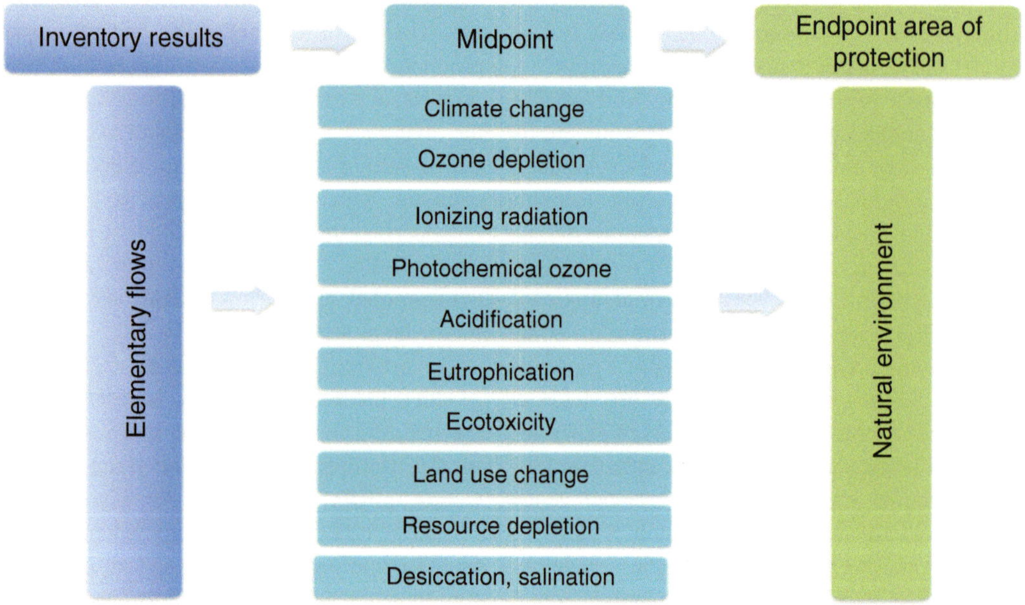

Fig. 6.2 Life Cycle Impact Assessment (LCIA) with environmental related factors, proposed by the Joint Research Centre

ally involve the use of high temperature and toxic solvents. Thus more sustainable alternative approaches should be developed to reduce the environmental load of the general production. The improved technologies for extraction are high-voltage electrical discharges, pulsed electric fields, and ultrasound treatments [5]. Yet these improved extraction technologies still require a significant input of energy and relatively high equipment cost.

The potential improvements that can lower the environmental burdens by the BacHBerry approach to produce high-added-value phenolics based on the proposed LCIA parameters are as follows:

- Less impacts on land-use change, owing to the biocatalytic processes which can be implemented all around the world. There is no need to build greenhouse in nontraditional berry cultivation areas to grow berries to produce phenolic compounds for food additives, nutraceuticals, or pharmaceuticals.
- Less impacts on resource depletion, implicating that even new compounds and new applications of phenolics would be discovered by

the screening platform of the project. The exploitation on this knowledge will not lead to overconsumption of the existing berry resources, due to the feedstock of the biocatalytic processes which would be those from common and variable biomass.

- Overall contribution to other factors listed in LCIA is because biocatalytic processes are in general regarded as more environmentally friendly compared to alternative chemical processes (e.g., using toxic chemical solvents to extract phenolic compounds from berries, as mentioned above).

Currently there is no biocatalytic process developed for phenolic compound production yet. However, the LCA on the productions of other bioactive products can provide a template for the possible processes developed by BacHBerry. LCA on beta-carotene extraction techniques is such an example [40], although the bioactive compound is obtained by chemical extraction. The LCA has been conducted on the extractions of beta-carotene from either carrots or microalgae. Life cycle inventory was built from production and each principal process for recovery of the compounds citing data from Ecoinvent 2.0 database. LCIA was also conducted using software Simapro 7.1 on cultivation, harvesting, drying, and yielding. The possible biocatalytic production of phenolic compounds would be similar to the extraction of beta-carotene from microalgae, while the conventional phenolic production approaches are similar to the one extracted from carrot. The case study of LCA on beta-carotene can therefore serve as a template to develop phenolic specific LCA.

6.4 Sustainable Assessment Based on Economic Impacts

Biocatalytic production of high-added-value phenolics holds great potential for sustainable development. To move from the laboratory to large-scale productions, these processes must pass a number of criteria to be implemented successfully. Other than safety, environmental, legal, and throughput issues, economic impacts are highly important [41]. Evaluating the cost of biocatalytic processes is difficult due to a lack of solid data, relevant case studies, and inventory on the factors contributing to the total cost. Comparing to the chemical manufacturing that has been studied in detail, the biocatalytic processes involve more a complex development chain than those in chemical manufacturing, making the economic assessments more difficult. Taking example from one extensively studied biocatalytic process, converting sugar to 1,3-propanediol, the biocatalytic process is economically competitive to the chemical process only at high cost of fossil feedstock plus when the sugar feedstock price is low [42]. Giving the similar scenery of BacHBerry approach to produce phenolic compounds via biocatalytic process, the cost of the biomass feedstock, as well as the other unforeseeable costs coming along the production chain, will have impact on their economic potential. The other cost includes cost on fermentation (scale, equipment, and yield), recovery, and purification. It is believed that more expensive products are worth producing in a higher catalyst cost scenario [41]. Thus the phenolic products for pharmaceutical applications may be the first ones viable to be produced via biocatalytic process based on the economic assessment.

The other impact on economy is the market and employment. The Centre for Strategy and Evaluation Services (CSES) estimated that bio-based products in general would contribute to increase in volume up to 38,000 million Euro in market growth from 2006 to 2020 and to create 260,000 more jobs [43]. Phenolic compounds produced via biocatalytic processes are among such bio-based products. Therefore they would also contribute in these aspects of the economy as well.

6.5 Sustainable Assessment Based on Social Impacts

Assessing social sustainability of biotech products is another critical aspect for the sustainable development. While environmental and economic indicators have a relatively broad consensus already, the social indicators remain in early stages of development.

To assess the social sustainability of biotech products, it shall be conducted taking into considerations the following:

- Make use of the scientific know-how for the sustainability assessment of biotechnological production.
- Develop a framework for the assessment of the social sustainability of biotech production within all stakeholders.
- Promote innovation toward sustainable development and the public engagement into these topics.

The BacHBerry project will build a broad-spectrum database on berries from around the world that provides a valuable scientific resource for future research. The project is a cooperation of research institutes, biotech, and science communication companies, which helps to build a platform for dialogue among the stakeholders and to engage the public alongside with the product development process.

The Dutch organization COGEM (Commissie Genetische Modificatie) has proposed how to assess the social sustainability of genetically modified (GM) crops while comparing to those grown by traditional agriculture [44]. The nine criteria brought up for GM crops could be applied to assess the benefit of biotech products to the society as well as shown in the table below.

Criteria	GM crops	BacHBerry-derived products
Benefit to society	Increase in yield, contributing to food security	Affordable quality products, similar or identical to the natural ones
Economics and prosperity	Efficiency of production process, productivity, and profit	Efficiency of production process, productivity, and profit
Health and welfare	Working environment, in terms of employment	Potential to improve human health and create new employment
Food supply	Food security, fair trade	Depending on the feedstock and scale of production
Cultural heritage	Offer room to conserve and continue specific cultural heritage aspects	Harnessing traditional knowledge and adding new knowledge associated with berries
Freedom of choice	Labeling of products, coexistence, research freedom	Maybe different from GM crops based on the final product formats
Safety	Food and environmental safety in accordance with national legislation and international agreements	Similar to the existing biotech products
Biodiversity	No damage or reduction to biodiversity	No damage or reduction to biodiversity
Environmental quality	Quality of soil, surface water and groundwater, and air does not deteriorate; greenhouse gas emission remains neutral	Full impacts will be evaluated based on the large-scale productions

6.6 Conclusion

Harnessing microbial production platform to produce high-added-value phenolic compounds has a wide range of applications across several industrial areas such as food (additives), functional food (nutraceuticals), and pharma (pharmaceuticals). The current process of the BacHBerry project toward developing suitable biocatalytic processes to produce phenolic compounds has been analyzed. The potential contributions of the general biocatalytic processes to sustainability have been evaluated in their environmental, economic, and social impacts, and they look promising. Once a biocatalytic process for a phenolic compound is finalized, a more detail assessment can then be conducted to show that biocatalytic processes are promising means to move toward sustainable development.

References

1. Mortensen A (2006) Carotenoids and other pigments as natural colorants. Pure Appl Chem 78:1477–1491
2. Atanasov AG, Waltenberger B, Pferschy-Wenzig EM, Linder T, Wawrosch C, Uhrin P, Temml V, Wang L, Schwaiger S, Heiss EH, Rollinger JM, Schuster D, Breuss JM, Bochkov V, Mihovilovic MD, Kopp B, Bauer R, Dirsch VM, Stuppner H (2015) Discovery and resupply of pharmacologically active plant-derived natural products: a review. Biotechnol Adv 33:1582–1614
3. Manach C, Scalbert A, Morand C, Remesy C, Jimenez L (2004) Polyphenols: food sources and bioavailability. Am J Clin Nutr 79:727–747
4. Mélanie G, Sophie C (2013) Polyphenol-rich beverages promote a sustainable and renewable generation of energy and prevent neurotoxicity. Int J Nutr Metab 5:28–39
5. Barba FJ, Galanakis CM, Esteve MJ, Frigola A, Vorobiev E (2015) Potential use of pulsed electric technologies and ultrasounds to improve the recovery of high-added value compounds from blackberries. J Food Eng 167:38–44
6. Basu A, Rhone M, Lyons TJ (2010) Berries: emerging impact on cardiovascular health. Nutr Rev 68:168–177
7. Mellway RD, Constabel CP (2009) Metabolic engineering and potential functions of proanthocyanidins in poplar. Plant Signal Behav 4:790–792
8. Patnaik R (2008) Engineering complex phenotypes in industrial strains. Biotechnol Prog 24:38–47
9. Peel GJ, Pang Y, Modolo LV, Dixon RA (2009) The LAP1 MYB transcription factor orchestrates anthocyanidin biosynthesis and glycosylation in Medicago. Plant J 59:136–149
10. Pscheidt B, Glieder A (2008) Yeast cell factories for fine chemical and API production. Microb Cell Fact 7:25
11. Xie DY, Sharma SB, Wright E, Wang ZY, Dixon RA (2006) Metabolic engineering of proanthocyanidins through co-expression of anthocyanidin reductase and the PAP1 MYB transcription factor. Plant J 45:895–907
12. Yu O, Jez JM (2008) Nature's assembly line: biosynthesis of simple phenylpropanoids and polyketides. Plant J 54:750–762
13. Invenire Market Intelligence (2008) Berries in the world. Available at http://www.sitra.fi/NR/rdonlyres/4A1F0F29-0B3C-458C-8843-D5436BEE6542/0/IMI08_Berriesintheworld.pdf
14. ESD:N. Basics of SD strategies. Available at http://www.sd-network.eu/?k=basics%20of%20SD%20strategies
15. OECD (2011) Draft OECD recommendation on assessing the sustainability of bio-based products. Available at http://www.oecd.org/sti/sci-tech/48222459.pdf
16. Longhi S, Giongo L, Buti M, Surbanovski N, Viola R, Velasco R, Ward JA, Sargent DJ (2014) Molecular genetics and genomics of the Rosoideae: state of the art and future perspectives. Hortic Res 1:1
17. Jardim C, Menezes R, Foito A, Stewart D, Ferreira RB, Santos CN (2014) European summer school on nutrigenomics. September 1–5, 2014 Camerino, Italy: abstracts. J Nutrigenet Nutrigenomics 7:75–93
18. Young KH (1998) Yeast two-hybrid: so many interactions, (in) so little time. Biol Reprod 58:302–311
19. Schreckinger ME, Lotton J, Lila MA, de Mejia EG (2010) Berries from South America: a comprehensive review on chemistry, health potential, and commercialization. J Med Food 13:233–246
20. Paredes-Lopez O, Cervantes-Ceja ML, Vigna-Perez M, Hernandez-Perez T (2010) Berries: improving human health and healthy aging, and promoting quality life – a review. Plant Foods Hum Nutr 65:299–308
21. Nakagawa A, Minami H, Kim JS, Koyanagi T, Katayama T, Sato F, Kumagai H (2011) A bacterial platform for fermentative production of plant alkaloids. Nat Commun 2:326
22. Boudet AM (2007) Evolution and current status of research in phenolic compounds. Phytochemistry 68:2722–2735
23. Yan Y, Chemler J, Huang L, Martens S, Koffas MA (2005) Metabolic engineering of anthocyanin biosynthesis in Escherichia coli. Appl Environ Microbiol 71:3617–3623
24. Nwafor CC, Gribaudo I, Schneider A, Wehrens R, Grando MS, Costantini L (2014) Transcriptome analysis during berry development provides insights into co-regulated and altered gene expression between a seeded wine grape variety and its seedless somatic variant. BMC Genomics 15, e1030
25. Donnez D, Jeandet P, Clement C, Courot E (2009) Bioproduction of resveratrol and stilbene derivatives by plant cells and microorganisms. Trends Biotechnol 27:706–713
26. Jang M, Cai L, Udeani GO, Slowing KV, Thomas CF, Beecher CW, Fong HH, Farnsworth NR, Kinghorn AD, Mehta RG, Moon RC, Pezzuto JM (1997) Cancer chemopreventive activity of resveratrol, a natural product derived from grapes. Science 275:218–220
27. Yang Y, Lin Y, Li L, Linhardt RJ, Yan Y (2015) Regulating malonyl-CoA metabolism via synthetic antisense RNAs for enhanced biosynthesis of natural products. Metab Eng 29:217–226
28. Li M, Kildegaard KR, Chen Y, Rodriguez A, Borodina I, Nielsen J (2015) De novo production of resveratrol from glucose or ethanol by engineered Saccharomyces cerevisiae. Metab Eng 32:1–11
29. Stahlhut SG, Siedler S, Malla S, Harrison SJ, Maury J, Neves AR, Forster J (2015) Assembly of a novel biosynthetic pathway for production of the plant fla-

vonoid fisetin in Escherichia coli. Metab Eng 31:84–93
30. Kallscheuer N, Vogt M, Kappelmann J, Krumbach K, Noack S, Bott M, Marienhagen J (2015) Identification of the phd gene cluster responsible for phenylpropanoid utilization in Corynebacterium glutamicum. Appl Microbiol Biotechnol 100:1871–1881
31. Wittmann C (2010) Analysis and engineering of metabolic pathway fluxes in Corynebacterium glutamicum. Adv Biochem Eng Biotechnol 120:21–49
32. Jegannathan KR, Nielsen PH (2013) Environmental assessment of enzyme use in industrial production – a literature review. J Clean Prod 42:228–240
33. ISO (2006) Environmental management -lifecycle assessment -requirements and guidelines (ISO 14044: 2006). In STANDARDIZATION ECF, ed. Brussels, Belgium
34. ISO (2006) Environmental management—life cycle assessment—principles and framework (ISO 14040: 2006)
35. JRC (2009) International Reference Life Cycle Reference Data System (ILCD) handbook. Available at http://eplca.jrc.ec.europa.eu/uploads/JRC-Reference-Report-ILCD-Handbook-Towards-more-sustainable-production-and-consumption-for-a-resource-efficient-Europe.pdf
36. JRC (2010) Framework and requirements for Life Cycle Impact Assessment (LCIA) models and indicators. Available at http://eplca.jrc.ec.europa.eu/uploads/ILCD-Handbook-LCIA-Framework-Requirements-ONLINE-March-2010-ISBN-fin-v1.0-EN.pdf
37. JRC (2010) Analysis of existing environmental impact assessment methodologies for use in Life Cycle Assessment (LCA). Available at http://eplca.jrc.ec.europa.eu/uploads/ILCD-Handbook-LCIA-Background-analysis-online-12March2010.pdf
38. JRC (2011) Recommendations for life cycle impact assessment in the European context. Available at http://eplca.jrc.ec.europa.eu/uploads/ILCD-Recommendation-of-methods-for-LCIA-def.pdf
39. Galanakis CM, Goulas V, Tsakona S, Manganaris GA, Gekas V (2013) A knowledge base for the recovery of natural phenols with different solvents. Int J Food Prop 16:382–396
40. Kyriakopoulou K, Papadaki S, Krokida M (2015) Life cycle analysis of β-carotene extraction techniques. J Food Eng 167:51–58
41. Tufvesson P, Lima-Ramos J, Nordblad M, Woodley JM (2011) Guidelines and cost analysis for catalyst production in biocatalytic processes. Org Process Res Dev 15:266–274
42. Hermann BG, Blok K, Patel MK (2007) Producing bio-based bulk chemicals using industrial biotechnology saves energy and combats climate change. Environ Sci Technol 41:7915–7921
43. CSES (2011) Final evaluation of the lead market initiative. Available at http://ec.europa.eu/enterprise/policies/innovation/policy/lead-market-initiative/final-eval_en.htm
44. COGEM (2009) Socio-economic aspects of GMOs. Available at http://ec.europa.eu/food/food/biotechnology/reports_studies/docs/Netherlands_annex_Cogem_report_en.pdf

Hindrances to the Efficient and Stable Expression of Transgenes in Plant Synthetic Biology Approaches

7

Ana Pérez-González and Elena Caro

Most agronomic traits and all metabolic pathways are controlled by multiple genes. Therefore, synthetic biology approaches that intend to recreate or modify them in plants require a multigene strategy. In complex approaches like these, where coordinated expression of multiple genes is required for stoichiometric synthesis of proteins or assembly of steps in a pathway, gene silencing is an especially worrisome problem since the instability of transgene expression can not only decrease the yield of production, but impair the whole functioning of the pathway. Thus, it is of vital importance to develop effective strategies for the generation of transgenic plants where uniform and predictable expression of transgenes can be achieved.

Since 1990, when Napoli, Lemieux, and Jorgensen first reported a silencing phenomenon [36], ample experimental data on loss of transgene expression has accumulated. The goal of their studies was to determine whether chalcone synthase (CHS), a key enzyme in flavonoid biosynthesis, was the rate-limiting enzyme in anthocyanin biosynthesis. The anthocyanin biosynthetic pathway is responsible for the violet coloration in petunias. In an attempt to generate deep violet petunias, Napoli and colleagues [36] overexpressed CHS, which unexpectedly resulted in white petunias. The levels of endogenous as well as introduced CHS were 50-fold lower than in wild-type petunias, which led them to hypothesize that the introduced transgene was "co-suppressing" the endogenous CHS gene. Twenty-five years later, it is clear that a way of tackling low transgene expression is to avoid epigenetic gene silencing in the transformed organism but we are still dealing with the design of strategies that successfully do it.

The silencing of transgenes results from the activation of defense mechanisms of the plant against foreign DNA [29, 30], a common occurrence in the stable integration of additional DNA into chromosomes (transposable elements (TEs)) and the replication of a viral genome (virus infection). Silencing can occur at the transcriptional level (transcriptional gene silencing (TGS)) either preventing or dampening transcription through DNA methylation and/or chromatin modifications, or at the posttranscriptional level (posttranscriptional gene silencing (PTGS)) through RNA cleavage or translational repression [27].

TGS is commonly associated with multiple and rearranged transgene copies and homology in promoter regions. It triggers cell-autonomous promoter hypermethylation and/or chromatin condensation that is maintained through mitosis

A. Pérez-González • E. Caro (✉)
Centro de Biotecnología y Genómica de Plantas, Universidad Politécnica de Madrid (UPM) - Instituto Nacional de Investigación y Tecnología Agraria y Alimentaria (INIA), Campus Montegancedo UPM, Pozuelo de Alarcón 28223, (Madrid), Spain
e-mail: elena.caro@upm.es

S. Singh (ed.), *Systems Biology Application in Synthetic Biology*,
DOI 10.1007/978-81-322-2809-7_7

and meiosis. PTGS is commonly associated with homology in coding regions transcribed from a strong promoter. It is believed to involve a threshold level of aberrant transcripts, triggering a sequence-specific RNA degradation mechanism that can spread through a phloem-transmissible signal. It can be accompanied by increased methylation in the corresponding transcribed DNA regions, but is typically reset through meiosis [29, 30, 44].

In any case, for the silencing to occur, small RNAs have to be generated from partially or perfectly double-stranded RNA (dsRNA) precursors by an RNase III-like nuclease called Dicer or Dicer-like (DCL). The small RNAs are incorporated into another nuclease named Argonaute (AGO), and they use Watson-Crick base pairing to guide the effector AGO complex to target nucleic acids [27] (Fig. 7.1).

The literature points to several factors in the generation of a transgenic plant that might be behind transgene licensing of silencing, mainly related to foreign DNA integration and organization within the host genome, the nature of its sequence, the regulatory elements controlling its expression, and its transcription. These factors, together with the most accepted strategies to minimize their effect, will be discussed in the following sections.

7.1 Genome Integration of Foreign DNA

It has been appreciated for many years that the structure of a transgenic locus and the state of the chromatin in the site of its integration can have a major influence on the level and stability of the transgene expression.

7.1.1 Structure of Transgenic Loci

Most genetic engineering of plants use *Agrobacterium*-mediated transformation to introduce novel genes. Although *Agrobacterium* mainly infects dicotyledonous plants in nature, it can genetically transform a wide range of higher plant species under laboratory conditions and has become the transformation vehicle of choice for the genetic manipulation of most plants [1, 9].

Monocotyledons were believed to be recalcitrant to transformation by *Agrobacterium tumefaciens*, but these initial difficulties have been eventually resolved, and all major cereals are now transformed quite efficiently by this method [16].

Direct insertion of naked DNA into plant cells is an alternative transformation strategy for all species, but it is especially useful for plants that are more difficult to transform using *Agrobacterium*. Among these methods, particle bombardment has become the most successful because it is based on purely mechanical principles and is therefore not dependent on the biological factors that restrict the *Agrobacterium* host range. Particle bombardment has been successfully applied to cereals including rice, maize, wheat, barley, and sorghum. Historically, sorghum was considered as one of the most recalcitrant major crops; however, transformation efficiency by particle bombardment has now improved from approximately 1 % to in excess of 20 % [25]. Other direct DNA transfer methods use chemicals (e.g., PEG, calcium phosphate) or physical treatments (e.g., electroporation) on plant protoplasts.

In all the mentioned cases, selection for antibiotic or herbicide resistance enables recovery of transformed cells that will then be regenerated to full transgenic plants.

Upon *Agrobacterium*-mediated transformation, usually intact, single or tandem T-DNA copies in one or two loci are stably integrated into AT-rich regions of the plant genome with minimal rearrangements of the target site. At low frequency, T-DNAs are truncated at their left border, and vector backbone DNA is integrated [1]. In contrast, direct DNA transfer often generates much larger transgenic loci, where high-copy numbers and extensive rearrangements of the foreign DNA have been frequently reported. The structure of such loci is highly variable, comprising single copies, tandem or inverted repeats, concatemers, intact transgenes, truncated and

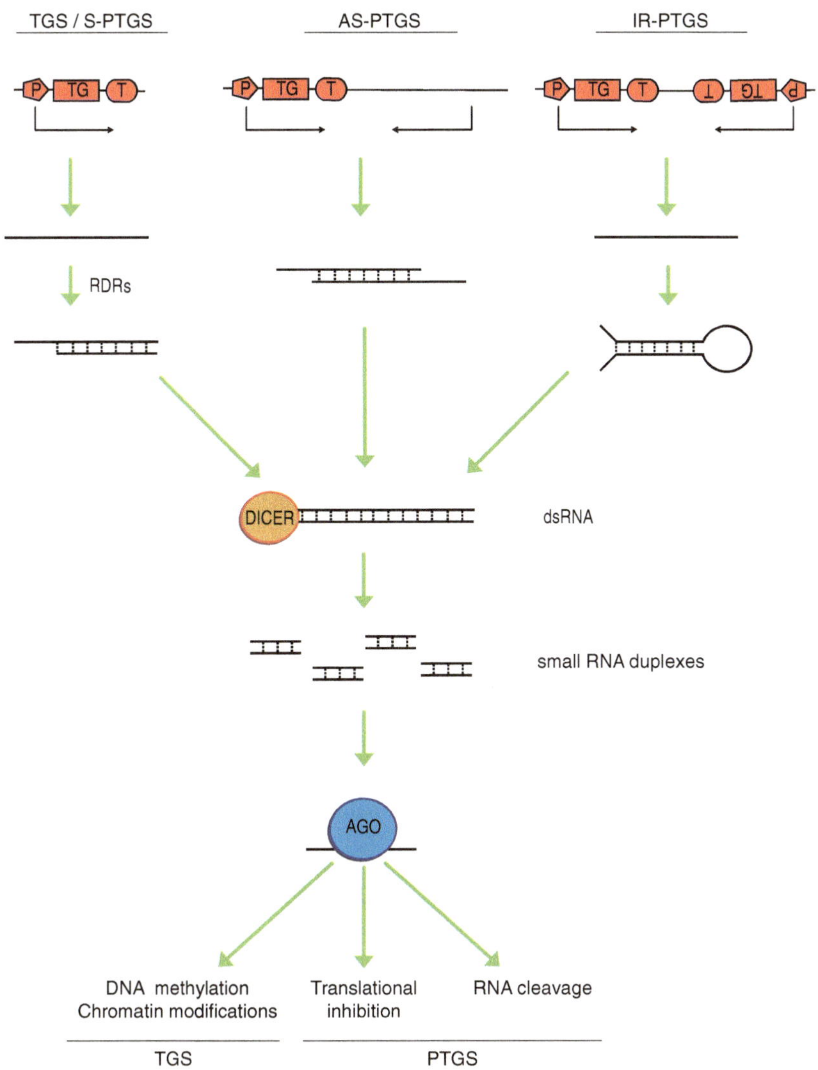

Fig. 7.1 Schematic representation of a model for RNA-based TGS and PTGS. TGS, triggered directly by single-copy transgenes through an unknown mechanism resulting in the methylation of their promoter region. S-PTGS (sense-PTGS), initiated by the generation of aberrant mRNAs by transgenes that will be the substrate for RDRs. AS-PTGS (antisense-PTGS), the consequence of the integration of a transgene next to an endogenous promoter leading to its antisense transcription. IR-PTGS (inverted repeat-PTGS), transcription of inverted copies of a transgene generating a hairpin RNA responsible for silencing. *P* promoter, *TG* transgene, *T* terminator. RDRs: RNA-dependent RNA polymerases, dsRNA: double-stranded RNA, DICER: endoribonucleases of the RNase III family that cleave dsRNA, AGO: family of Argonaute proteins that bind small RNAs and coordinate downstream gene-silencing events guided to their targets by sequence complementarity

rearranged sequences, and interspersed genomic DNA [1, 20].

The existence of repeat-sensitive transcriptional repression mechanisms, described long ago in plants and animals, establishes that single gene copies at a defined locus are expressed much more effectively than reiterated transgenes [49]. Thus, there seems to be a consensus in the field that to avoid silencing, an *Agrobacterium*-based delivery method should be favored for the introduction of foreign genes into plants, together with the selection of transgenic lines that show a single-site insertion with a single copy of the intact transgene or transgenes [1] (Fig. 7.2).

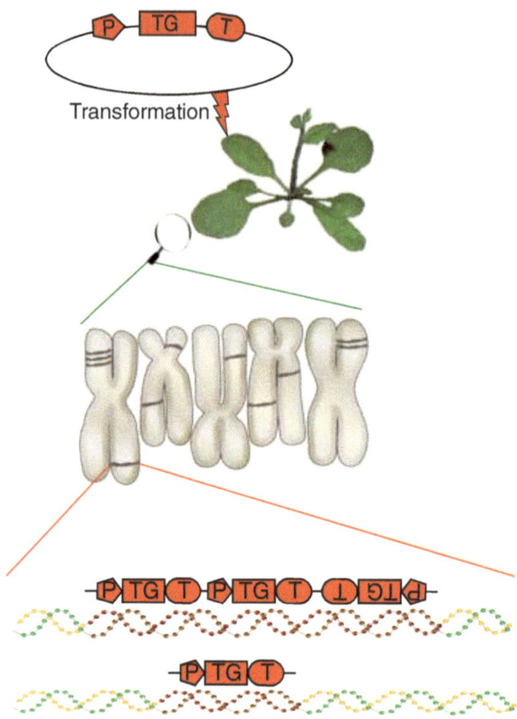

Fig. 7.2 The different methods of integration of transgenes in the genome of a plant can lead to very different situations. Multiple insertion sites or multiple copies inserted at a site often lead to silencing of the transgenes. Single insertion of single-copy genes is the preferred situation in the search for transgenics with efficient and stable expression. *P* promoter, *TG* transgene, *T* terminator

7.1.2 Positional Effect

An important cause of interindividual variability during plant transformation experiments is the chromosomal position effect that arises in response to the site within the genome into which the foreign transgenic DNA has integrated [29].

Previous work from numerous laboratories has suggested that integration of *Agrobacterium tumefaciens* T-DNA into the plant genome occurs preferentially in promoter or transcriptionally active regions. However, under nonselective conditions, a relatively high frequency of T-DNA insertions have been found in heterochromatic regions, including centromeres, telomeres, and rDNA repeats. It is possible that recovery of T-DNAs in these regions is disfavored under selective conditions because the insertion of the selection marker in heterochromatin ends up with a loss of expression of the transgene [18].

Additionally, positional effect affects transgenes that are integrated near endogenous regulatory elements, such as transcriptional enhancers or repressors, which can cause their misexpression.

Several strategies that can be followed to avoid these problems, like targeted integration of transgenes and the use of locus control regions, which will be presented in detail.

7.1.2.1 Targeted Integration

One possible approach to address positional effect is to precisely integrate a single copy of the transgene of interest into a predefined target locus that is characterized by long-term stable expression.

For a long time, it was not possible to use double-strand break (DSB) induction for gene targeting due to the lack of means to direct DSBs to specific sites, but in the last years, there has been a huge development of genome-editing techniques based on the generation of modified nucleases and synthetic DNA-targeting strategies. Domains derived from zinc-finger transcription factors or transcription activator-like effectors have been used to design modules that recognize a DNA sequence of choice. The fusion of these modules to an endonuclease domain can now introduce DSBs at the selected specific sites [39]. The recently discovered CRISPR/Cas system based on RNA-guided engineered nucleases is yet a new tool to induce multiple DSBs that holds great promise due to its simplicity, efficiency, and versatility [3].

The insertion of the transgenic constructs from a donor vector at the selected loci where DSBs have been produced would, ideally, allow for high-level transcription and isolation from endogenous regulatory elements. The use of site-specific nucleases could, moreover, remove much of the regulatory burden associated with transgenic plants since one of the main causes of concern to the regulatory authorities is the random integration of transgenes and the resulting potential for unintended effects such as disrupting host metabolism and/or producing toxic or allergenic compounds [3].

These strategies for gene targeting have already proven successful [38], although they are still at an early stage. Recipient lines with characterized "safe harbor" loci promoting the strong expression of transgenes still have to be established, and methods for selection need to be optimized until they become routinely used.

7.1.2.2 Use of Locus Control Regions

Random integration of transgenes can interfere with resident gene function and the endogenous gene expression regulation program and as a result have its own expression affected as well. Various mechanisms exist within eukaryotic genomes to avoid enhancer-mediated activation of nearby promoters and chromosomal position effects [17]. Transgenic constructs lack this ability and thus require supplementary ways to minimize such disturbances.

Genetic insulators are sequences that function to shield genes from outside signals preventing inappropriate activation or repression of expression by nearby regulatory elements. Possibly one of the most well-studied class of genetic insulators is scaffold/matrix attachment regions (S/MARs), which have been suggested to function as boundary elements, anchoring the ends of chromosomal domains and preventing the spreading of heterochromatin into transgenes flanked by them [2] (Fig. 7.3). Early experiments in *Arabidopsis* did not show a clear effect on transgene expression by the use of S/MARs [43], however, since then many groups have reported that their use causes an increase in the level of transgene expression and/or a reduction in plant-to-plant variability in different species, including *Arabidopsis* [41].

A few years ago, Kishimoto and colleagues [19] reflected on the fact that some transgenes undergo TGS while others do not, making it conceivable that there are endogenous DNA sequences that actively determine the epigenetic TGS/non-TGS state of genomic regions. They developed a screening strategy to identify such elements (which they called anti-silencing regions (ASRs)), based on their ability to protect a flanked transgene from TGS. They succeeded in identifying three ASRs from *Lotus japonicus* that included Ty1/copia retrotransposon-like and pararetrovirus-like sequences. They could show that one retrotransposon-like sequence had interspecies anti-TGS activity in *Arabidopsis thaliana*, and it held a lot of promise due to its small size (171 bp) that would make it very convenient to include in the flanks of any transgenic construct.

7.2 Transgene Sequence Composition

In the genome, most genes are present in isochores covering an extremely narrow GC range of 1–2 %, suggesting that any exogenous DNA with different features might be detected as intrusive. In fact, TEs, prokaryotic sequences, GA-rich microsatellites, retroelement remnants, and tandem repeat arrays are the primary elements correlated with silencing [22]. The different

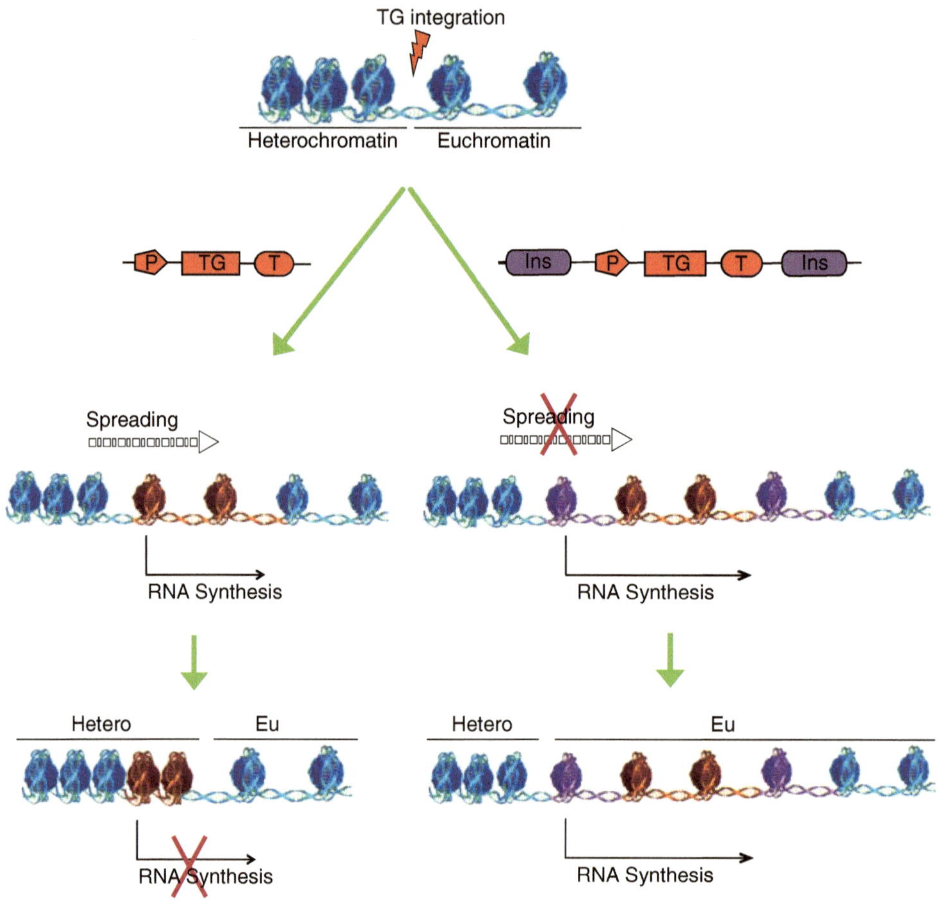

Fig. 7.3 Genetic insulators can shield transgenes from outside signals preventing positional effects caused by heterochromatin spreading from the integration site in the genome. *P* promoter, *TG* transgene, *T* terminator, *Ins* genetic insulator, *Hetero* heterochromatin, *Eu* euchromatin

sensitivities to methylation of a monocotyledonous versus a dicotyledonous transgene in petunia [11, 32] suggested long ago that silencing can be provoked by particular sequence contexts. Prokaryotic DNA might be recognized as foreign because of its generally high GC content and/or because it cannot be packaged properly with eukaryotic proteins [29].

To avoid alerting plant genome surveillance mechanisms as a defense against intrusive foreign DNAs, modification of transgenic construct sequences should be made as necessary to make sure that all element sequences match isochore composition of the host species [48].

7.3 Promoter and Terminator Usage

Throughout plant development, small RNAs target homologous genomic DNA sequences for cytosine methylation in all sequence contexts through TGS via the phenomenon termed RNA-directed DNA methylation (RdDM) [23] (Fig. 7.1). RdDM has been proven responsible for the *de novo* initiation, reestablishment, and maintenance of TEs and transgene silencing. In this last case, silencing is commonly associated with a specific increase in DNA methylation within the promoter region [31].

Small RNAs direct the molecular machinery that catalyzes heterochromatic histone modifications or DNA methylation to loci with sequence homology, usually by base pairing with noncoding RNAs (ncRNAs) that are associated with the chromatin at the locus to be silenced. Thus, a low level of transcripts needs to be generated to provide positional information for TGS. RNA Polymerase IV is believed to produce single-stranded RNAs that serve as precursors of small RNAs. RNA Polymerases V and II, in contrast, are involved in producing the ncRNA scaffolds with which 24 nucleotide small RNAs form base pairs [14].

There are some known players involved in the recruitment of Pol IV and Pol V to target sequences like transposons and repeats that already carry epigenetic silenced features. However, the pathway leading to the initiation of the silencing process in the case of transgene promoters remains elusive [14, 31].

The genome-wide high-resolution mapping and functional analysis of DNA methylation in *Arabidopsis* revealed that only about 5% of genes contain methylation within promoter regions [50]. Whether this resistance of endogenous promoters to silencing is based on their structure, sequence or any other feature is not known and remains to be elucidated.

Using constitutive viral promoters with very different sequence features to those of the host genome has repeatedly shown not to be a good approach to achieve high and stable transgene expression. As an example, the 35S promoter of the Cauliflower mosaic virus has been documented in many instances and different species to end up silenced and methylated (Table 7.1). The promoters chosen to drive transgene expression are essential regulatory elements that often get overlooked, and further work on this matter will be necessary to find the best-suited candidates for each experiment.

Table 7.1 Examples of species of transgenic plants where DNA methylation of the 35S Cauliflower mosaic virus promoter was reported

Reference	Species common name	Species scientific name
Weber and Graessmann [46]	Tobacco	*Nicotiana tabacum*
Meyer et al. [33]	Petunia	*Petunia hybrida*
Kumar and Fladung [21]	Aspen	*Populus tremula*
Chalfun-Junior et al. [4]	*Arabidopsis*	*Arabidopsis thaliana*
Mishiba et al. [34]	Gentian	*Gentiana triflora* × *G. scabra*
Gambino et al. [13]	Grapevine	*Vitis* spp.
Sohn et al. [42]	–	*Nicotiana benthamiana*
Fan et al. [12]	Sweet orange	*Citrus sinensis Osb.*
Weinhold et al. [47]	–	*Nicotiana attenuata*
Okumura et al. [37]	Lettuce	*Lactuca sativa*

7.4 Transgene Transcription

Besides the small RNA pathways that regulate endogenous genes and transposons, plants have developed a small RNA pathway dedicated mainly to the control of viruses. It is also often activated against transgenes expressed under the control of strong promoters (S-PTGS) as a consequence of the saturation of the mRNA processing pathways [24] (Fig. 7.1). This saturation translates in the accumulation of aberrant RNAs that are converted into dsRNA by RDRs. A plausible scenario is that cap-, poly (A)- and other RNA-binding proteins normally prevent RDRs from interacting with mRNAs. In misprocessed RNAs with aberrant characteristics, these RNA-binding proteins would bind inefficiently allowing the generation of dsRNA by RDRs [35].

However, highly transcribed endogenes, for example, the ribulose-1,5-bisphosphate carboxylase/oxygenase (RuBisCO) gene, are not silenced. Transgene RNAs can be expected to be particularly prone to aberrancy if they have non-plant-derived elements, because they may not have the precise structures necessary for efficient interaction with the mRNA-binding proteins associated with most cellular mRNAs [15]. This observation suggests that qualitative rather than quantitative features of transcripts define whether silencing is initiated or not [8].

Given that introns are very common in endogenous genes but are often lacking in transgenes and transposons, it was hypothesized that introns may suppress gene silencing. This idea is supported by results showing that three different introns from *Arabidopsis* genes increase the expression of GFP when introduced in its 5′UTR [6]. In fact, an endogene-resembling transgene (which was modified to include two introns) showed a delay in the onset of silencing compared to its intronless version [8], and several proteins of both the splicing and the polyadenylation machineries have been identified as regulators of DNA methylation patterns and chromatin silencing [28].

In IR-PTGS (Fig. 7.1), dsRNA generated from the transcription of inverted repeats efficiently silences the corresponding transgene mRNA. This can be the result of a deliberate design of the construct to generate dsRNAs and induce silencing, or the consequence of the integration of inverted copies on the genome. PTGS can also be initiated by antisense transcription of the transgene (AS-PTGS; Fig. 7.1), deliberately, as a means to induce silencing, or as the consequence of the integration of the transgene in the genome next to an endogenous promoter leading to its antisense transcription. Once again, the selection of transgenic lines with single-copy insertions and with no transgene rearrangements and the use of genetic insulators flanking transgenes are of the utmost importance to avoid positional effects.

For transient expression approaches, the strategies used to solve PTGS problems consist on the co-expression of the gene of interest with a viral silencing suppressor. So far, several suppressors of RNA silencing have been identified that seem to interfere with the PTGS silencing pathway at distinct steps, affecting various molecular targets in the host. Researchers have used the Artichoke mottled crinkle virus suppressor P19 in *Agrobacterium* infiltration transient expression assays to produce high yields of biopharmaceuticals, namely, a human antibody against the tumor-associated antigen tenascin-C in *N. tabacum* [45] and the HIV-1 Nef protein in *N. benthamiana* [7].

But the use of viral suppressors is not a good solution to the overall problem. On the first hand, they have been found to work in a dose-dependent manner that can be easily controlled in the lab for transient expression assays, but not in stably transformed plants, where the high doses have been shown to yield plants with deformed phenotypes, for example, in the case of expression of P19 in *A. thaliana* [10], *N. tabacum* [5], and *N. benthamiana* [40]. This can be due to the fact that the tampering with silencing mechanisms also affects the normal expression of endogenous genes necessary for a correct development. Moreover, many of the most potent suppressors are pathogenicity factors that often contribute to the onset of symptoms upon infection of plants.

7.5 Strategies to Avoid Transgene Silencing

Synthetic biology complex approaches involving the transfer of multiple genes into plants absolutely require stable transgene expression to be successful. As described in the above sections, there are some strategies that should be followed to increase the probabilities of achieving it, and we will summarize them here.

Selecting a method of DNA delivery that minimizes the number of copy inserts within the host genome and the screening for transgenic lines with no transgene rearrangements is important to obtain stable lines with consistent expression through many generations.

In the near future, it will be possible to avoid the positional effect derived from the integration site by choosing between a handful of euchromatic sites within the genome to integrate your transgene of interest, but as of now, if random integration methods are used, several lines should be followed in case some suffer from spreading of heterochromatin neighboring the transgene. In any case, it will always be advisable to flank the transgenic cassettes with genetic isolators that can somehow shelter the DNA from changes in the surroundings and from AS-PTGS that could derive from integration next to an antisense promoter.

It is advisable for the transgene to match the isochore AT/GC composition of the host organism genome and that plasmid sequences must be excluded from the integrated DNA to avoid foreign DNA recognition.

The choice of promoters and terminators is also important in the design of the transgenic construct. Until a thorough analysis of regulatory sequences' features that induce silencing is made, the use of viral sequences or of artificial sequences with very different AT/CG contents from the host genome average should in general be avoided. It might also be interesting to design different alternatives with promoters and terminators of varying strengths in order to not saturate the RNA maturation machinery.

In the case of multigene approaches, a common question is whether it is advisable to use the same promoter and terminator sequences repeatedly to control the expression of multiple genes. In theory, the use of diverse elements to build up the transcriptional units should be preferred in order to avoid repetition and initiation of TGS.

It must be noted that there are examples in the literature of successful experiments in which co-expression of multiple genes has been achieved with repetitious promoters [26], especially in the field of metabolic engineering [51]. However, as synthetic biology initiatives become more ambitious, the current strategy of selecting for the best performing lines and discarding the many others in which the expression of transgenes does not behave as expected must be improved. We propose that the design of strategies that take into account all the above mentioned issues will increase the rate of success of future endeavors.

Much work is still needed to elucidate the different signals that lead to the generation of dsRNAs from transgenes, to understand the stochasticity of the phenomena and the specifics of how the pathway works in each different species, but until then, taking all these precautions to avoid gene silencing might make the difference between success and failure in a synthetic biology approach.

References

1. Anami S, Njuguna E, Coussens G et al (2013) Higher plant transformation: principles and molecular tools. Int J Dev Biol 57:483–494. doi:10.1387/ijdb.130232mv
2. Bode J, Benham C, Knopp A, Mielke C (2000) Transcriptional augmentation: modulation of gene expression by scaffold/matrix-attached regions (S/MAR elements). Crit Rev Eukaryot Gene Expr 10:73–90
3. Bortesi L, Fischer R (2014) The CRISPR/Cas9 system for plant genome editing and beyond. Biotechnol Adv 33:41–52. doi:10.1016/j.biotechadv.2014.12.006
4. Chalfun-Junior A, Mes JJ, Mlynárová L et al (2003) Low frequency of T-DNA based activation tagging in Arabidopsis is correlated with methylation of CaMV 35S enhancer sequences. FEBS Lett 555:459–463. doi:10.1016/S0014-5793(03)01300-0
5. Chiba M, Reed JC, Prokhnevsky AI et al (2006) Diverse suppressors of RNA silencing enhance agro-infection by a viral replicon. Virology 346:7–14. doi:10.1016/j.virol.2005.09.068
6. Christie M, Croft LJ, Carroll BJ (2011) Intron splicing suppresses RNA silencing in Arabidopsis. Plant J 68:159–167. doi:10.1111/j.1365-313X.2011.04676.x
7. Circelli P, Donini M, Villani ME et al (2010) Efficient Agrobacterium-based transient expression system for the production of biopharmaceuticals in plants. Bioeng Bugs 1:221–224. doi:10.4161/bbug.1.3.11722
8. Dadami E, Moser M, Zwiebel M et al (2013) An endogene-resembling transgene delays the onset of silencing and limits siRNA accumulation. FEBS Lett 587:706–710. doi:10.1016/j.febslet.2013.01.045
9. De Cleene M, De Ley J (1976) The host range of crown gall. Bot Rev 42:389–466. doi:10.1007/BF02860827
10. Dunoyer P, Lecellier C-H, Parizotto EA et al (2004) Probing the microRNA and small interfering RNA pathways with virus-encoded suppressors of RNA silencing. Plant Cell 16:1235–1250. doi:10.1105/tpc.020719
11. Elomaa P, Helariutta Y, Griesbach RJ et al (1995) Transgene inactivation in Petunia hybrida is influenced by the properties of the foreign gene. Mol Gen Genet 248:649–656
12. Fan J, Liu X, Xu S-X et al (2011) T-DNA direct repeat and 35S promoter methylation affect transgene expression but do not cause silencing in transgenic sweet orange. Plant Cell Tissue Organ Cult 107:225–232. doi:10.1007/s11240-011-9973-z
13. Gambino G, Perrone I, Carra A et al (2009) Transgene silencing in grapevines transformed with GFLV resistance genes: analysis of variable expression of transgene, siRNAs production and cytosine methylation. Transgenic Res 19:17–27. doi:10.1007/s11248-009-9289-5

14. He X-J, Ma Z-Y, Liu Z-W (2014) Non-coding RNA transcription and RNA-directed DNA methylation in Arabidopsis. Mol Plant 7:1406–1414. doi:10.1093/mp/ssu075
15. Herr AJ, Molnar A, Jones A, Baulcombe DC (2006) Defective RNA processing enhances RNA silencing and influences flowering of Arabidopsis. Proc Natl Acad Sci 103:14994–15001. doi:10.1073/pnas.0606536103
16. Hiei Y, Ishida Y, Komari T (2014) Progress of cereal transformation technology mediated by Agrobacterium tumefaciens. Front Plant Sci 5:628. doi:10.3389/fpls.2014.00628
17. Kadauke S, Blobel GA (2009) Chromatin loops in gene regulation. Biochim Biophys Acta 1789:17–25. doi:10.1016/j.bbagrm.2008.07.002
18. Kim S-I, Gelvin SB (2007) Genome-wide analysis of Agrobacterium T-DNA integration sites in the Arabidopsis genome generated under non-selective conditions. Plant J 51:779–791. doi:10.1111/j.1365-313X.2007.03183.x
19. Kishimoto N, Nagai J, Kinoshita T et al (2013) DNA elements reducing transcriptional gene silencing revealed by a novel screening strategy. PLoS One 8, e54670. doi:10.1371/journal.pone.0054670
20. Kohli A, Twyman RM, Abranches R et al (2003) Transgene integration, organization and interaction in plants. Plant Mol Biol 52:247–258
21. Kumar S, Fladung M (2000) Determination of transgene repeat formation and promoter methylation in transgenic plants. Biotechniques 28:1128 1130, 1132, 1134 passim
22. Kumpatla SP, Chandrasekharan MB, Iyer LM et al (1998) Genome intruder scanning and modulation systems and transgene silencing. Trends Plant Sci 3:97–104. doi:10.1016/S1360-1385(97)01194-1
23. Law JA, Jacobsen SE (2010) Establishing, maintaining and modifying DNA methylation patterns in plants and animals. Nat Rev Genet 11:204–220. doi:10.1038/nrg2719
24. Lindbo JA, Silva-Rosales L, Proebsting WM, Dougherty WG (1993) Induction of a highly specific antiviral state in transgenic plants: implications for regulation of gene expression and virus resistance. Plant Cell 5:1749–1759. doi:10.1105/tpc.5.12.1749
25. Liu G, Campbell BC, Godwin ID (2014) Sorghum genetic transformation by particle bombardment. Methods Mol Biol 1099:219–234. doi:10.1007/978-1-62703-715-0_18
26. Liu W, Stewart CN (2015) Plant synthetic biology. Trends Plant Sci 20:309–317. doi:10.1016/j.tplants.2015.02.004
27. Martínez de Alba AE, Elvira-Matelot E, Vaucheret H (2013) Gene silencing in plants: a diversity of pathways. Biochim Biophys Acta 1829:1300–1308. doi:10.1016/j.bbagrm.2013.10.005
28. Mathieu O, Bouché N (2014) Interplay between chromatin and RNA processing. Curr Opin Plant Biol 18:60–65. doi:10.1016/j.pbi.2014.02.006
29. Matzke AJ, Matzke MA (1998) Position effects and epigenetic silencing of plant transgenes. Curr Opin Plant Biol 1:142–148
30. Matzke MA, Matzke AJ (1998) Epigenetic silencing of plant transgenes as a consequence of diverse cellular defence responses. Cell Mol Life Sci 54:94–103
31. Matzke MA, Mosher RA (2014) RNA-directed DNA methylation: an epigenetic pathway of increasing complexity. Nat Rev Genet 15:394–408. doi:10.1038/nrg3683
32. Meyer P, Heidmann I (1994) Epigenetic variants of a transgenic petunia line show hypermethylation in transgene DNA: an indication for specific recognition of foreign DNA in transgenic plants. Mol Gen Genet 243:390–399
33. Meyer P, Linn F, Heidmann I et al (1992) Endogenous and environmental factors influence 35S promoter methylation of a maize A1 gene construct in transgenic petunia and its colour phenotype. MGG Mol Gen Genet 231:345–352. doi:10.1007/BF00292701
34. Mishiba K, Nishihara M, Nakatsuka T et al (2005) Consistent transcriptional silencing of 35S-driven transgenes in gentian. Plant J 44:541–556. doi:10.1111/j.1365-313X.2005.02556.x
35. Moreno AB, Martínez de Alba AE, Bardou F et al (2013) Cytoplasmic and nuclear quality control and turnover of single-stranded RNA modulate post-transcriptional gene silencing in plants. Nucleic Acids Res 41:4699–4708. doi:10.1093/nar/gkt152
36. Napoli C, Lemieux C, Jorgensen R (1990) Introduction of a chimeric chalcone synthase gene into Petunia results in reversible co-suppression of homologous genes in trans. Plant Cell 2:279–289. doi:10.1105/tpc.2.4.279
37. Okumura A, Shimada A, Yamasaki S et al (2015) CaMV-35S promoter sequence-specific DNA methylation in lettuce. Plant Cell Rep. doi:10.1007/s00299-015-1865-y
38. Petolino JF, Kumar S (2015) Transgenic trait deployment using designed nucleases. Plant Biotechnol J. doi:10.1111/pbi.12457
39. Puchta H, Fauser F (2014) Synthetic nucleases for genome engineering in plants: prospects for a bright future. Plant J 78:727–741. doi:10.1111/tpj.12338
40. Siddiqui SA, Sarmiento C, Truve E et al (2008) Phenotypes and functional effects caused by various viral RNA silencing suppressors in transgenic Nicotiana benthamiana and N. tabacum. Mol Plant Microbe Interact 21:178–187. doi:10.1094/MPMI-21-2-0178
41. Singer SD, Liu Z, Cox KD (2012) Minimizing the unpredictability of transgene expression in plants: the role of genetic insulators. Plant Cell Rep 31:13–25. doi:10.1007/s00299-011-1167-y
42. Sohn S-H, Choi MS, Kim K-H, Lomonossoff G (2011) The epigenetic phenotypes in transgenic Nicotiana benthamiana for CaMV 35S-GFP are mediated by spontaneous transgene silencing. Plant Biotechnol Rep 5:273–281. doi:10.1007/s11816-011-0182-3

43. Thompson WF, Spiker S, Allen GC (2007) Regulation of transcription in eukaryotes. In: Grasser KD (ed) Regulation of transcription in plants. Wiley-Blackwell, Oxford
44. Vaucheret H, Béclin C, Elmayan T et al (1998) Transgene-induced gene silencing in plants. Plant J 16:651–659
45. Villani ME, Morgun B, Brunetti P et al (2009) Plant pharming of a full-sized, tumour-targeting antibody using different expression strategies. Plant Biotechnol J 7:59–72. doi:10.1111/j.1467-7652.2008.00371.x
46. Weber H, Graessmann A (1989) Biological activity of hemimethylated and single-stranded DNA after direct gene transfer into tobacco protoplasts. FEBS Lett 253:163–166. doi:10.1016/0014-5793(89)80951-2
47. Weinhold A, Kallenbach M, Baldwin IT (2013) Progressive 35S promoter methylation increases rapidly during vegetative development in transgenic Nicotiana attenuata plants. BMC Plant Biol 13:99. doi:10.1186/1471-2229-13-99
48. De Wilde C (2000) Plants as bioreactors for protein production: avoiding the problem of transgene silencing. In: Matzke MA, Matzke AJM (eds) Plant gene silencing. Springer, Dordrecht
49. Wolffe AP (1997) Transcription control: repressed repeats express themselves. Curr Biol 7:R796–R798. doi:10.1016/S0960-9822(06)00408-8
50. Zhang X, Yazaki J, Sundaresan A et al (2006) Genome-wide high-resolution mapping and functional analysis of DNA methylation in arabidopsis. Cell 126:1189–1201. doi:10.1016/j.cell.2006.08.003
51. Zorrilla-López U, Masip G, Arjó G et al (2013) Engineering metabolic pathways in plants by multigene transformation. Int J Dev Biol 57:565–576. doi:10.1387/ijdb.130162pc

8 The New Massive Data: miRnomics and Its Application to Therapeutics

Mohammad Ahmed Khan, Maryam Mahfooz, Ghufrana Abdus Sami, Hashim AlSalmi, Abdullah E.A. Mathkoor, Ghazi A. Damanhauri, Mahmood Rasool, and Mohammad Sarwar Jamal

8.1 Introduction

Genomic medicine is highly dependent on understanding the biological processes regulating gene expression. In this reference, the discovery of phenomena of RNA interference in 1998 served as turning point for the field of genomic medicine. It was observed that the double stranded RNA (dsRNA) is capable of silencing specific genes in *Caenorhabditis elegans* [12]. Later, studies on RNA interference have revealed that RNA interference operates in many species and serves in silencing genes. In *C. elegans*, the inhibitory potential of RNA was induced by introducing endogeneous long dsRNA's while in mammalian cells, the introduction of small 21 nt RNA's could induce RNAi. [10].

Among the small RNAs, small noncoding RNAs (sncRNAs) form the most dominant class of RNAs [22]. Human gene expression is regulated through small noncoding RNAs (sncRNAs) in a very precise manner. MicroRNA (miRNA) is one such endogenous sncRNA which is involved in the negative regulation of gene expression. It inhibits the translation or causes the degradation of RNA by binding to the 3′ UTR of the target RNA [40]. The effect depends on whether the complementation is imperfect (inhibition of translation) or perfect (degradation) [11]. As a group, miRNAs regulate more than 50 % of protein coding genes which accounts for more than 10,000 genes.

miRNAs are involved in cell differentiation, proliferation/growth, mobility apoptosis and many other cellular functions. These cellular effects of miRNAs are seen in multiple tissue types [4, 24, 25, 32]. miRNA's thus play key roles in several physiological and developmental processes. Considering the importance of miRNAs, it is not unanticipated that miRNA are also in turn regulated in a stringent manner. Evidence suggests that any alteration of miRNA regulation can lead to diseases such as cancer, heart disease, hepatic disorder, metabolic and immune dysfunctions. Since miRNA regulate multiple proteins and pathways, their importance in next generation therapeutics can be envisioned.

M.A. Khan
National Institute of Biologicals, Noida, UP, India

M. Mahfooz
Department of Computer Science, Jamia Millia Islamia, New Delhi 110025, Delhi, India

G.A. Sami • A.E.A. Mathkoor
Department of Biotechnology, Jamia Millia Islamia, New Delhi 110025, Delhi, India

H. AlSalmi • G.A. Damanhauri • M.S. Jamal, Ph.D. (✉)
King Fahd Medical Research Center (KFMRC), King Abdulaziz University, Jeddah, Saudi Arabia
e-mail: sarwar4u@gmail.com

M. Rasool
Center of Excellence in Genomic Medicine Research (CEGMR), King Abdulaziz University, Jeddah, Saudi Arabia

S. Singh (ed.), *Systems Biology Application in Synthetic Biology*,
DOI 10.1007/978-81-322-2809-7_8

"microRNomics" therefore has emerged as a field of human disease biology and a subdiscipline of genomics for studying the expression, biogenesis and regulation of expression of several target proteins. It is therefore essential to understand the biological functions of miRNA' on a genomic scale.

Misregulation of miRNAs is associated with the development of many diseases [39]. Therefore miRNAs have been receiving special importance in the field of drug design [37, 49]. Both miRNA replacement therapy and specific miRNA inhibitors are being tried on for the restoration of normal tissue functions [17, 33]. In order to enhance the endogenous level of specific miRNAs, miRNA mimetics can be used. miRNAs can also suppress the expression of genes involved in disease progression [27]. These mimetic or inhibitory actions on miRNA regulated processes have shown promising therapeutic response [31].

8.2 miRNA-Based Therapeutic Strategies

There exist a lot of similarities between the development of miRNA-based therapeutics and the conventional drug discovery process. However, unlike the conventional drug discovery process, selection of miRNAs targets is based on preexisting knowledge since miRNA are endogenous molecules with well-defined regulatory functions [14, 43, 48]. The primary step would therefore be the identification of dysregulated miRNAs in a particular disease followed by selection of the candidate miRNA. This miRNA is then functionally characterized using suitable *in vitro* and *in vivo* experiments to quantify the gain or loss of function. Based on the gain or loss, either replacement or inhibitory strategies are developed (Fig. 8.1).

8.2.1 miRNA Inhibition

Over expressed miRNAs levels are often the cause of several diseases. In such cases, the prevention or reversal of miRNA expression has been found beneficial. For example, increased level of miR-122 has been implicated in hepatitis C where overexpression of miR-122 favors parasite replication [18]. This is evident from studies which show that upon miR-122 inhibition, the viral load is reduced [19]. Over expression of miR-21 [36] is also a cause of several cancers. Over expression of miR-21 causes increased cell proliferation through cell cycle alterations. Similarly, overexpression of miR-212/132 is observed in pathological hypertrophy of heart [41]. Since numerous miRNAs are reported to be

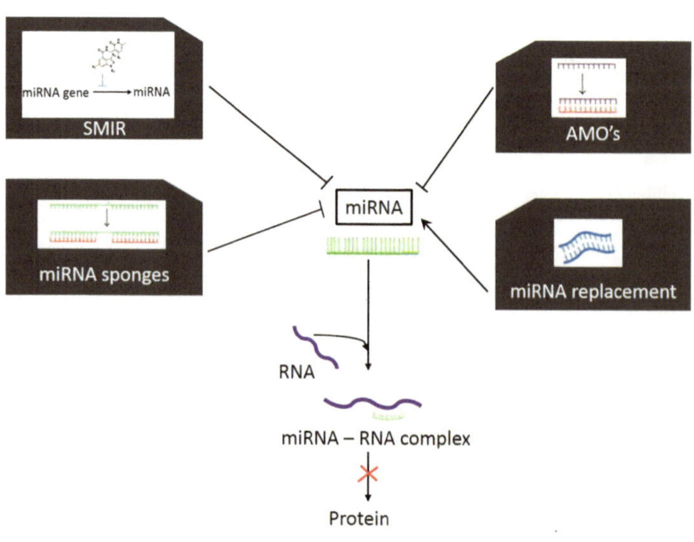

Fig. 8.1 miRNA-based therapeutic strategies for enhancing or repressing miRNA functions

overexpressed in many different diseases, miRNA inhibition has also become a major research area of in the field of gene therapy.

8.2.1.1 Methods for miRNA Inhibition

miRNA Sponges

The method of using miRNA "sponge" was introduced to induce continuous loss of function of miRNA in cell lines and transgenic organisms. Sponge RNAs are a series of miRNA response elements which contain complementary binding sites to a miRNA of interest. miRNA sponges occur naturally in plants and animals as long non-coding RNA. Like majority of miRNA target genes, sponge also inhibits a whole family of miRNA as the sponge's binding site is situated in the seed region of miRNA. As many cells (both *in vitro* and *in vivo*) are resistance to the uptake of oligonucleotides, the sponge transgene is usually delivered by a viral vector. Sponge mRNAs are usually designed synthetically and are either viral vectors or plasmids having upto 10 arrayed miRNA binding sites with small nucleotide spacers [8, 9].

MiRNA sponges have been well studied against hepatocellular carcinoma and in other cancer types. Recently, a lentivirus mediated sponge for microRNA-122 targeting cyclin G1, Bcl-w, disintegrin and metalloprotease 10 has been developed. microRNA-122 plays an important silencing role in the Huh7 hepatoma cell line and the U2OS osteocarcinoma cell line. miR-122-SP can efficiently restore the expression of miR-122. Moreover, miR-122 sponge was effective in suppression of proliferation through cell cycle arrest at G1 phase and activation of caspase-3/7 in both hepatoma and osteosarcoma cells [26]. Circular miRNA sponges have also been developed for miR-21 or miR-221 which showed excellent anticancer effect against malignant melanoma cells. These, miRNA, being circularized, are less susceptible to enzymatic degradation while being immune to miRNA-mediated degradation. It also had superior efficacy in depressing microRNA targets vis-a-vis linear sponges and other inhibitors [23]. The miR-101 is a negative regulator of amyloid precursor protein (of amyloid β which is responsible for neurodegeneration in Alzheimer's disease). A lentiviral sponge for miR-101 is reported to regulate the amyloid precursor protein metabolism in hippocampal neurons. This indicated miR-101 inhibition can control the amyloidogenic processing signifying its importance in the Alzheimer's disease [1].

Anti-miRNA Oligonucleotides (AMO)

Anti-miRNA Oligonucleotides (AMO) are synthetic oligonucleotides (19–25 nt long) which work on the principle of antisense techniques to intervene with the target miRNA [47]. The earliest report of miRNA inhibition using AMOs was observed in Drosophila embryos [2].

AMOs are reverse complements of miRNA which work by inducing steric blockage with their respective miRNA. AMOs either degrade the miRNA through their RNase activity or prevent its binding to the target mRNA. The most important properties of AMOs are they have high binding affinity and specificity. It also has scope for chemical modifications which can help in improving its potency as well as performance [21]. First generation AMOs have 2′-O-methyl modifications which are termed as antagomirs. 2′-O-methyl modification ensures that AMOs are resistant to nucleases also facilitate miRNA binding. Second generation AMOs were modified at the 2′ sugar position to provide better nuclease resistance and improved binding affinity compared to first generation [20]. Locked nucleic acid (LNA) modifications are characterized by bicyclic nucleic acid having methylene bridge. LNAs have shown better binding affinity; however, in some cases, this higher affinity has also resulted in off-target binding leading to toxicity [38]. Some AMOs have lot of chemical modifications and are reportedly good at inhibiting non-coding as well as coding RNAs [16]. The potential of using AMOs for clinical applications is increasing. Anti-miR-122 oligonucleotides have shown promising therapeutic potential against chronic hepatitis C virus in the long-term safety and efficacy trials [42]. An LNA-modified oligonucleotide is reported to potently inhibit cardiomyocyte-specific miR-208a function leading to suppression of fibrosis, diminished expression of myosin 7 and improved survival of Dahl salt-sensitive rats having diastolic dysfunction when on high salt diet [30].

Currently, AMOs are most researched area for developing miRNA therapeutics. Targeting of multiple miRNAs using single fragment, termed as multiple-target AMO technology (MT-AMO), has also emerged in last 2–3 years. This technology allows use of single AMO fragment having 2′-O-methyl-modified oligoribonucleotides to target multiple miRNA;s or miRNA seed families [46]. After the regulatory approval of first generation oligonucleotide Vitravene for CMV retinitis, the potential for modified AMOs is on the rise, especially in the area of cancer biology. Fully modified oligonucleotides such as 20-mer phosphorodiamidate morpholino oligomer targeting c-Myc are currently being investigated in human trials [7]. OMe-oligonucleotides and mixed backbone OMe/DNA hybrid antisense oligonucleotides are current being pursued to correct aberrant splicing events [28]. The focus is therefore on the practical usage of miRNAs to try and find out cure for various diseases.

Small Molecular Inhibitors of Specific miRNAs (SMIR)

Melo and Calin et al. were first to use the term small-molecule drugs targeting specific miRNAs (SMIR) to identify interaction of small molecules and miRNAs. SMIR approach has promising potential in modulation of miRNA activity. It can overcome the developmental challenges posed with nucleotide analogs. The SMIR-approach can reduce the duration of drug development, making it cost effective. It can help in development of more targeted therapies [29, 50]. An azobenzene was discovered as the first specific SMIR against miR-21 precursor [15]. Current approach in SMIR involves identification of compounds with potent and specific binding affinity towards mature miRNAs or its upstream precursor. In this sense, small molecules would be targeting a mature miRNA sequence by binding to it, or to any of its upstream precursors. Ongoing research envisages identifying small molecules with structural complementarities to miRNAs showing structure based interaction. However, the major limitation in the development of SMIR is that not many crystal structures of miRNAs are reported. Also the use of SMIRs is limited due to their high EC50 values. However SMIRs are relatively easy to deliver. Despite the limitations, bench to bedside delivery of SMIRs is comparatively easier. Aryl amides have been recently discovered as a new class of SMIR that serves as an inhibitor of miR-21, which is frequently upregulated in cardiac diseases and cancers [5].

8.2.2 miRNA Replacement Therapy

Till now, the research on therapeutic approaches with miRNA has mostly focused on inhibition of miRNA. However, miRNA replacement therapy has also emerged with a proof of concept. As the name suggests, miRNA replacement therapy aims to restore the healthy state by increasing the amount of miRNAs [34]. The best examples are let-7 [3] and miR-34 [6] which are tumor suppressors whose reduced levels have been characterized in many tumor types. Similarly, decrease in miR-107 is characterized in early stages of Alzheimer's disease making it a promising target for replacement therapy [45]. MiRNA mimics can inhibit the genes targeted by suppressor miRNAs and consequently normalize cellular processes. It is important that miRNA mimics are delivered through targeted approach to prevent miRNA over expression beyond basal level and to bypass normal tissues. Mimics of miRNA also serve as an attractive substrate for nucleases mediated degradation. The data on miRNA replacement therapy suggests that some diseases like cancer manifest impaired miRNA processing which leads to global miRNA downregulation. Therefore, for such cases an agent which can upregulate the expression of a particular miRNAs is needed [35]. The area of miRNA replacement therapy is growing slowly; however, a miR-34 mimic currently under clinical trial for treatment of solid tumors has shown the silver lining.

8.3 Future Prospects

The reported *in vitro* and *in vivo* studies on miRNA inhibitors and inhibition of miRNAs support further research on miRNAs as lead compounds. While the cost of drug development is increasing day by day with the regulatory requirements becoming more stringent, it becomes essential that a drug candidate must be identified quickly and validated properly. As miRNAs are short, the primary screening of an ideal candidate against the miRNA must account for an in-depth understanding of the specificity. There are many challenges in the field of miRNomics. Reported miRNA inhibitions only focus on the target tissues and a little emphasis is laid on the possible off-target effects. AntimiR development is based on the principle that targeting any particular miRNA will regulate all genes under it. It should be noted that miRNAs also target other unrelated genes which may possibly produce unwanted or undesired alterations in gene expression. For example, miR208a was studied for its cardiac effects but it also showed anti-obesity behavior and was active against metabolic syndrome in mice [13, 44]. In addition, many times therapeutically non-feasible doses have been reported and separate studies to develop dosage regimen will be essential. The miRNAs also partly share their targets, thus interaction of a particular miRNA with a target weakens its potency for interaction with other targets. On the other hand, interaction of one mRNA with a specific miRNA reduces the probability of its silencing by other miRNA. Much more exploration is yet to be carried out in this area of physiological competition. Therefore simultaneous targeting of multiple pathways using combinatorial approaches of multiple miRNAs could be more effective strategy while reducing cost of therapy.

Despite these challenges, targeting of miRNA using mimics or inhibitors is now established as a realistic option against many human diseases. Many of these synthetically developed miRNAs have reached the clinical stage as mentioned above and it is expected that even higher number will be approved for testing at clinical stage in coming years. However, for achieving success, continued research and exploration of miRNAs as a new class of drug targets is the need of the hour.

References

1. Barbato C, Pezzola S, Caggiano C, Antonelli M, Frisone P, Ciotti MT, Ruberti F (2014) A lentiviral sponge for miR-101 regulates RanBP9 expression and amyloid precursor protein metabolism in hippocampal neurons. Front Cell Neurosci 8:37
2. Boutla A, Delidakis C, Tabler M (2003) Developmental defects by antisense-mediated inactivation of micro-RNAs 2 and 13 in *Drosophila* and the identification of putative target genes. Nucleic Acids Res 31:4973–4980
3. Boyerinas B, Park SM, Hau A, Murmann AE, Peter ME (2010) The role of let-7 in cell differentiation and cancer. Endocr Relat Cancer 17(1):F19–F36
4. Bravo-Egana V, Rosero S, Klein D, Jiang Z, Vargas N, Tsinoremas N et al (2012) Inflammation-mediated regulation of MicroRNA expression in transplanted pancreatic islets. J transplant 2012:723614
5. Connelly CM, Deiters A (2014) Identification of inhibitors of microRNA function from small molecule screens. Methods Mol Biol 1095:147–156
6. Craig VJ, Cogliatti SB, Imig J, Renner C, Neuenschwander S, Rehrauer H, Schlapbach R, Dirnhofer S, Tzankov A, Müller A (2011) Myc-mediated repression of microRNA-34a promotes high-grade transformation of B-cell lymphoma by dysregulation of FoxP1. Blood 117(23):6227–6236
7. Devi GR, Beer TM, Corless CL, Arora V, Weller DL, Iversen PL (2005) *In vivo* bioavailability and pharmacokinetics of a c-MYC antisense phosphorodiamidate morpholino oligomer, AVI-4126, in solid tumors. Clin Cancer Res 11:3930–3938
8. Ebert MS, Neilson JR, Sharp PA (2007) MicroRNA sponges: competitive inhibitors of small RNAs in mammalian cells. Nat Methods 4(9):721–726
9. Ebert MS, Sharp PA (2010) MicroRNA sponges: progress and possibilities. RNA 16(11):2043–2050
10. Elbashir SM, Harborth J, Lendeckel W, Yalcin A, Weber K, Tuschl T (2001) Duplexes of 21-nucleotide RNAs mediate RNA interference in cultured mammalian cells. Nature 411(6836):494–498
11. Filipowicz W, Bhattacharyya SN, Sonenberg N (2008) Mechanisms of post-transcriptional regulation by microRNAs: are the answers in sight? Nat Rev Genet 9(2):102–114
12. Fire A, Xu S, Montgomery MK, Kostas SA, Driver SE, Mello CC (1998) Potent and specific genetic interference by double-stranded RNA in *Caenorhabditis elegans*. Nature 391(6669):806–811
13. Grueter CE, van Rooij E, Johnson BA, DeLeon SM, Sutherland LB, Qi X, Gautron L, Elmquist JK, Bassel-Duby R, Olson EN (2012) A cardiac microRNA

governs systemic energy homeostasis by regulation of MED13. Cell 149(3):671–683
14. Guessous F, Zhang Y, Kofman A, Catania A, Li Y, Schiff D, Purow B, Abounader R (2010) MicroRNA-34a is tumor suppressive in brain tumors and glioma stem cells. Cell Cycle 9(6):1031–1036
15. Gumireddy K, Young DD, Xiong X, Hogenesch JB, Huang Q, Deiters A (2008) Small-molecule inhibitors of microrna miR-21 function. Angew Chem Int Ed Engl 47(39):7482–7484
16. Henry SP, Geary RS, Yu R, Levin AA (2001) Drug properties of second-generation antisense oligonucleotides: how do they measure up to their predecessors? Curr Opin Investig Drugs 2:1444–1449
17. Janet B, Amir H (2008) Therapy analysis-microRNA; update analysis. Pharmaproject, 29
18. Jopling CL, Yi M, Lancaster AM, Lemon SM, Sarnow P (2005) Modulation of hepatitis C virus RNA abundance by a liver-specific MicroRNA. Science 309(5740):1577–1581
19. Lanford RE, Hildebrandt-Eriksen ES, Petri A, Persson R, Lindow M, Munk ME, Kauppinen S, Ørum H (2010) Therapeutic silencing of microRNA-122 in primates with chronic hepatitis C virus infection. Science 327(5962):198–201
20. Lennox KA, Behlke MA (2010) A direct comparison of anti-microRNA oligonucleotide potency. Pharm Res 27(9):1788–1799
21. Lennox KA, Behlke MA (2011) Chemical modification and design of anti-miRNA oligonucleotides. Gene Ther 18(12):1111–1120
22. Li Z, Rana TM (2014) Therapeutic targeting of microRNAs: current status and future challenges. Nat Rev Drug Discov 13(8):622–638
23. Liu Y, Cui H, Wang W, Li L, Wang Z, Yang S, Zhang X (2013) Construction of circular miRNA sponges targeting miR-21 or miR-221 and demonstration of their excellent anticancer effects on malignant melanoma cells. Int J Biochem Cell Biol 45(11):2643–2650
24. Long J, Wang Y, Wang W, Chang BH, Danesh FR (2011) MicroRNA-29c is a signature microRNA under high glucose conditions that targets Sprouty homolog 1, and its in vivo knockdown prevents progression of diabetic nephropathy. J Biol Chem 286:11837–11848
25. Lynn FC, Skewes-Cox P, Kosaka Y, McManus MT, Harfe BD, German MS (2007) MicroRNA expression is required for pancreatic islet cell genesis in the mouse. Diabetes 56:2938–2945
26. Ma J, Wu Q, Zhang Y, Li J, Yu Y, Pan Q, Sun F (2014) MicroRNA sponge blocks the tumor-suppressing functions of microRNA-122 in human hepatoma and osteosarcoma cells. Oncol Rep 32(6):2744–2752
27. Mack GS (2007) MicroRNA gets down to business. Nat Biotechnol 25(6):631–638
28. Mani S, Goel S, Nesterova M, Martin RM, Grindel JM, Rothenberg ML et al (2003) Clinical studies in patients with solid tumors using a second-generation antisense oligonucleotide (GEM 231) targeted against protein kinase A type I. Ann N Y Acad Sci 1002:252–262
29. Melo S, Villanueva A, Moutinho C, Davalos V, Spizzo R, Ivan C, Rossi S, Setien F, Casanovas O, Simo-Riudalbas L, Carmona J, Carrere J, Vidal A, Aytes A, Puertas S, Ropero S, Kalluri R, Croce CM, Calin GA, Esteller M (2011) Small molecule enoxacin is a cancer-specific growth inhibitor that acts by enhancing TAR RNA-binding protein 2-mediated microRNA processing. Proc Natl Acad Sci U S A 108:4394–4399
30. Montgomery RL, Hullinger TG, Semus HM, Dickinson BA, Seto AG, Lynch JM, Stack C, Latimer PA, Olson EN, van Rooij E (2011) Therapeutic inhibition of miR-208a improves cardiac function and survival during heart failure. Circulation 124(14):1537–1547
31. Ohlsson Teague EM, Van der Hoek KH, Van der Hoek MB, Perry N, Wagaarachchi P, Robertson SA, Print CG, Hull LM (2009) MicroRNA-regulated pathways associated with endometriosis. Mol Endocrinol 23(2):265–275
32. Pullen TJ, da Silva Xavier G, Kelsey G, Rutter GA (2011) miR-29a and miR-29b contribute to pancreatic beta-cell-specific silencing of monocarboxylate transporter 1 (Mct1). Mol Cell Biol 31:3182–3194
33. Raver-Shapira N, Marciano E, Meiri E, Spector Y, Rosenfeld N, Moskovits N, Bentwich Z, Oren M (2007) Transcriptional activation of miR-34a contributes to p53-mediated apoptosis. Mol Cell 26(5):731–743
34. Roush S, Slack FJ (2008) The let–7 family of microRNAs. Trends Cell Biol 18:505–516
35. Scott GK, Mattie MD, Berger CE, Benz SC, Benz CC (2006) Rapid alteration of microRNA levels by histone deacetylase inhibition. Cancer Res 66:1277–1281
36. Sicard F, Gayral M, Lulka H, Buscail L, Cordelier P (2013) Targeting miR-21 for the therapy of pancreatic cancer. Mol Ther 21(5):986–994
37. Simonson B, Das S (2015) MicroRNA therapeutics: the next magic bullet? Mini Rev Med Chem 15(6):467–474
38. Swayze EE, Siwkowski AM, Wancewicz EV, Migawa MT, Wyrzykiewicz TK, Hung G, Monia BP, Bennett CF (2007) Antisense oligonucleotides containing locked nucleic acid improve potency but cause significant hepatotoxicity in animals. Nucleic Acids Res 35(2):687–700
39. Glaser V (2008) Tapping miRNA-regulated pathways. Genet Eng Biotechnol News 28(5)
40. Tay Y, Kats L, Salmena L, Weiss D, Tan SM, Ala U, Karreth F, Poliseno L, Provero P, Di Cunto F, Lieberman J, Rigoutsos I, Pandolfi PP (2011) Coding-independent regulation of the tumor suppressor PTEN by competing endogenous mRNAs. Cell 147(2):344–357
41. Ucar A, Gupta SK, Fiedler J, Erikci E, Kardasinski M, Batkai S, Dangwal S, Kumarswamy R, Bang C,

Holzmann A, Remke J, Caprio M, Jentzsch C, Engelhardt S, Geisendorf S, Glas C, Hofmann TG, Nessling M, Richter K, Schiffer M, Carrier L, Napp LC, Bauersachs J, Chowdhury K, Thum T (2012) The miRNA-212/132 family regulates both cardiac hypertrophy and cardiomyocyte autophagy. Nat Commun 3:1078

42. van der Ree MH, van der Meer AJ, de Bruijne J, Maan R, van Vliet A, Welzel TM, Zeuzem S, Lawitz EJ, Rodriguez-Torres M, Kupcova V, Wiercinska-Drapalo A, Hodges MR, Janssen HL, Reesink HW (2014) Long-term safety and efficacy of microRNA-targeted therapy in chronic hepatitis C patients. Antivir Res 111:53–59
43. van Rooij E, Purcell AL, Levin AA (2012) Developing microRNA therapeutics. Circ Res 110(3):496–507
44. van Rooij E, Quiat D, Johnson BA, Sutherland LB, Qi X, Richardson JA, Kelm RJ Jr, Olson EN (2009) A family of microRNAs encoded by myosin genes governs myosin expression and muscle performance. Dev Cell 17(5):662–673
45. Wang WX, Rajeev BW, Stromberg AJ, Ren N, Tang G, Huang Q, Rigoutsos I, Nelson PT (2008) The expression of microRNA miR-107 decreases early in Alzheimer's disease and may accelerate disease progression through regulation of beta-site amyloid precursor protein-cleaving enzyme 1. J Neurosci 28(5):1213–1223
46. Wang Z (2011) The concept of multiple-target anti-miRNA antisense oligonucleotide technology. Methods Mol Biol 676:51–57
47. Weiler J, Hunziker J, Hall J (2006) Anti-miRNA oligonucleotides (AMOs): ammunition to target miRNAs implicated in human disease? Gene Ther 13(6):496–502
48. Yan LX, Wu QN, Zhang Y, Li YY, Liao DZ, Hou JH, Fu J, Zeng MS, Yun JP, Wu QL, Zeng YX, Shao JY (2011) Knockdown of miR-21 in human breast cancer cell lines inhibits proliferation, in vitro migration and in vivo tumor growth. Breast Cancer Res 13(1):R2
49. Zhang C (2008) MicroRNomics: a newly emerging approach for disease biology. Physiol Genomics 33(2):139–147
50. Zhang S, Chen L, Jung EJ, Calin GA (2010) Targeting microRNAs with small molecules: from dream to reality. Clin Pharmacol Ther 87:754–758

9 Microscopy-Based High-Throughput Analysis of Cells Interacting with Nanostructures

Raimo Hartmann and Wolfgang J. Parak

Nowadays, nanotechnology is everywhere. Engineered nanomaterials can be found in everyday products but also in cutting-edge technology. Since the mid-1980s, when the term "nanoparticle" (NP) first appeared in the context it is used nowadays, a "new" branch of science emerged. This direction of research has its roots in classical disciplines, in particular colloidal chemistry. The new interest originated for several reasons. First, new tools were developed which allowed the systematic organization and manipulation of matter on the nanometer length scale. Second, ideas were developed on how to apply nanoparticles in other disciplines, in particular for biological labeling and for photovoltaics. Today, nanomaterials are in the focus of research in several disciplines, with a much wider focus, including the application in molecular biology and medicine, but also in catalysis and energy conversion/storage [1–3].

Being reduced to several nanometers, the physicochemical properties of matter change. This can be related to the following aspects: (i) Surface-dependent properties of the bulk material such as chemical reactivity, soil-repellant features, or surface conductivity are becoming more dominant due to the dramatically increased surface-to-volume ratio. (ii) Size-dependent effects become visible and detectable, for instance, as superparamagnetism. (iii) Quantum mechanical properties are altered, which can result in new optical characteristics, for example, size-dependent changes in the absorption/emission spectra [1, 4, 5].

Apart from interesting physicochemical features for material sciences, nanomaterials bear some interesting properties for biomedical applications. They are small enough to be internalized by eukaryotic cells and can be targeted by surface modifications or external stimuli to some degree [6–8]. Superparamagnetic nanoparticles (e.g., from iron oxide) and plasmonic nanoparticles (e.g., from gold) can both be applied for hyperthermia, though due to different underlying phenomena [9–12]. With magnetic nanoparticles, energy from alternating magnetic fields is converted into heat, while plasmonic NPs convert UV/visible light into heat. Apart from that, luminescent NPs, such as quantum dots (QDs), are suitable for labeling or tracking purposes in molecular biology and medical diagnosis. This is due to their excellent optical characteristics, such as narrow emission/excitation bands and high photostability [13–15]. In addition, nanoparticles are utilized for intracellular sensing and delivery [16–18], and researchers are trying to target diseases such as cancer or Alzheimer's disease [19–21].

Although nanoparticles are already applied in vivo since the early 1990s, the interactions

This chapter is adopted from the PhD thesis of Dr. Raimo Hartmann, submitted and accepted by the Physics Department of the Philipps Universität Marburg in 2015.

R. Hartmann (✉) • W.J. Parak
Fachbereich Physik, Philipps Universität Marburg, Marburg, Germany
e-mail: raimo.hartmann@physik.uni-marburg.de

S. Singh (ed.), *Systems Biology Application in Synthetic Biology*,
DOI 10.1007/978-81-322-2809-7_9

with biological systems are so far not entirely understood on the single cell level. During the last decade, huge efforts were spent to unravel the dependency between endocytic uptake and several parameters of the nanoparticulate material, such as size [22, 23], shape [24], surface charge/chemistry [25–28], or stiffness [29–31] in vitro. Although this led to an improvement in the general understanding, most studies are lacking comparability, as the experimental conditions are extremely diverse. This applies to the selection of cells, the exposure conditions, different assay endpoints, or low significance of the studies carried out. Also, difficulties are the result of the almost continuous creation of more and more different nanomaterials and the fact that the area of bionanotechnology is very interdisciplinary [18].

Generally, once suspended in biological fluids, proteins and other biomolecules are adsorbed on the nanoparticle surface forming a layer called biomolecular (protein) corona. It is assumed that the biological identity of the NP and the interaction with cells are largely defined by this corona [32–34]. Upon cellular internalization, which typically happens through energy-dependent endocytic pathways, NPs are mostly transported to lysosomes (degradative intercellular organelles), where they are enriched [27, 28, 35, 36]. Regarding in vitro experiments, lysosomal accumulation is often accompanied by increasing cytotoxicity [28, 37]. In animal models, accumulation of NPs was observed mainly in the liver, spleen, and kidneys [38–40].

Cytometry describes the measurement of cell properties. Nanoparticle-cell interactions are commonly studied with microscopy-based methods. The method of determining characteristics of cells from microscope images is referred to as image cytometry. Many nanomaterials are intrinsically fluorescent or are designed to be functionalized easily with fluorescent dyes. Hence, fluorescence microscopy or variants of this method are typically used for imaging. Biological systems can have complex architecture, but the building blocks, i.e., individual cells, appear to be rather similar, as the same substructures (i.e., nucleus, outer plasma membrane, cytoskeleton, mitochondria, certain vesicles, etc.) can be found inside most of them. All of these unique substructures have unique properties (e.g., specific architecture or certain constituents). Based on these properties, they can be recognized within virtually any cell and thus, if stained and imaged appropriately, in any image representing a cell. Therefore, a visual model can be created, which describes how a cell, which was treated with certain dyes or exhibits certain fluorescent patterns, typically appears on a micrograph. Based on this model, a computer is now able to "see" and identify any cell being similar to the proposed model, including its constituents. As a result, the examination process can be automated. The computer-aided process of assessing cell properties is referred to as digital image cytometry in the following. This kind of image analysis is not reflecting the subjective perception of the experimenter any more. Additionally, the analysis process is much faster and the number of analyzable cells is dramatically increased, together with the statistical significance of the obtained results.

This principle of digital image cytometry is utilized in high-content analysis (HCA).[1] HCA is used to describe the screening and examination of thousands of cells ("content") in microscope images generated usually by automated microscopes in high throughput. HCA is mostly applied in biotechnological research, drug discovery, and in the workflow of pharmaceutical industry. It is either used to identify substances that trigger desired cellular responses or for assessing cytotoxicity in vitro [41–43]. Generally, it is regarded as a "multiparametric interrogation of cellular processes in any format" [41]. Important research fields where HCA-based assays were employed are, for instance, neurobiology [44, 45], oncology [46, 47], cell signaling [48, 49], or target identification and validation [50, 51].

In basic research in the field of nanobiotechnology, multiparametric response and cytotoxicity studies are needed to be able to fully correlate cell functions with the parameters of the deployed nanomaterial in a systematic manner. Remarkably, such questions can often be answered with one

[1] Also referred to as high-content screening (HCS).

HCA-based approach by multiplexing different assays with fluorescent probes spread across the visible spectrum [41]. In addition, this knowledge may help to estimate health and environmental hazards upon disposal of and exposure to certain potentially toxic nanomaterials.

So far, several published studies can be found utilizing HCA in a broader context for assessing nanoparticle-cell interactions: An extensive work about the cytotoxicity of cationic and anionic amine-modified polystyrene NPs ($d_h \approx 50$ nm)[2] including seven different cell lines[3] was performed by Anguissola et al. The analysis of the HCA data revealed that for cationic NPs, first (in terms of lowest concentration of NPs) lysosomal alkalinization occurs, which is followed by the loss of mitochondrial membrane potential, nuclear condensation, the increase of cytosolic calcium levels, and finally the disturbance of the integrity of plasma membranes. The effects were observed in a certain order but at similar concentrations, where viability (in terms of cell count) was decreased. For anionic particles, these effects could not be observed in the investigated range of NP-concentrations [28].

Similarly, but less well-performed, cellular responses were assessed upon exposure to L-cysteine-stabilized Au NPs[4] and 3 nm-sized cadmium telluride (CdTe) quantum dots using HCA by Jan et al. Interestingly, cellular proliferation and mitochondrial membrane potential were already reduced at concentrations almost two orders of magnitude lower ($\approx$1 nM) than those where acute cytotoxicity was observed (>50 nM) in terms of reduced cell count and loss of plasma membrane integrity [52].

The cellular effects of gold nanoparticles were also investigated by Soenen et al. They reported that exposure to poly(isobutylene-alt-maleic anhydride)-graft-dodecyl-coated NPs of 4 nm in core diameter and concentrations above 50 nM reduced cellular viability, cell size, cell proliferation, and differentiation in endothelial cells. Additionally, neurite outgrowth was impeded in neural progenitor cells. Furthermore, deformations in the actin and tubulin cytoskeleton were observed [53].

Solmesky et al. utilized HCA for studying the toxicity of lipid-based nanoparticles ($d_h \approx 100$ nm) in fibroblasts depending on the nanoparticle surface charge at physiological pH [54]. Several parameters were assessed including viability, proliferation, and morphological changes of mitochondria. Cationic nanoparticles turned out to be the most cytotoxic in terms of cell viability, which is also in line with previous findings [25, 55]. In addition, a decrease in mitochondrial elongation was observed [54].

These studies were selected because the benefits of the application of HCA, and thus the benefit of digital image cytometry, are comprehensively demonstrated. Especially in the first article, the authors could reconstruct the cellular mechanisms, which eventually lead to cell death upon exposure to nanoparticles, in a very systematic manner [28].

In digital image cytometry, measurements of cell properties are derived from microscopic images (in 2D or 3D) by applying algorithms. This approach is closely linked to computer vision, as the automatic recognition (segmentation) of individual cells is required. All cellular features with unique morphometric, densiometric, or textural properties can be investigated provided that their imaging is possible [56, 57]. In combination with high-throughput microscopy, valuable datasets containing profiles of thousands of individual cells can be obtained within a short time. Digital image cytometry is the basis for high-content analysis, which is used in biological research and drug discovery, to identify substances altering the cellular phenotype in a desired manner [41, 43].

Comparing the results from image cytometry with classical flow cytometry/imaging flow cytometry [58], discrepancies are present when applying the two techniques to similar cell samples [59, 60]. In the case of flow cytometry, fluorescence-labeled cells pass a laser beam one by one. From the momentary pulse of emitted photons caused by single cell-crossing events,

[2] d_h = hydrodynamic diameter.

[3] 1321 N1, SH-SY5Y, Raw267.4, A549, hCMEC, HepG2, and HEK293 cells.

[4] Jan et al. [52] did not provide any further characterization in their work.

the amount of fluorescence can be used to correlate the labeling efficiency with cellular functioning. Differences may be caused by the fact that the imaging conditions are completely different and thus, the results have to be carefully normalized to be comparable in absolute values.

The main advantage of Imaging Cytometry is the usage of digital microscopy and therefore the capability to "look into the cell" with high spatial resolution. Hundreds of parameters can be quantified which are not accessible by classical flow cytometry. Finally, the capability to analyze time-lapse image data lends itself to observations of the evolution of certain parameters over time, by following individual cells during movement or tracking particles during cellular uptake [41, 43].

9.1 Requirements

Digital image cytometry is typically performed on image sets gathered by fluorescence microscopy, as this microscopy technique allows for visualizing specific structures exclusively. However, for automated computer vision, it is mandatory to additionally obtain information of features of the cellular framework for cell identification (Image Segmentation, Sect. 9.4).

In conventional absorption light microscopy, image contrast is generated by an inhomogeneous absorption or scattering profile of the specimen, which can be altered by introducing dyes. Distinct structures are only capable of being differentiated in case they bear different optical properties. All colors are usually registered in the same image. In contrast, in fluorescence microscopy, structures of interest are fluorescently labeled with dyes emitting at different wavelengths upon excitation. The fluorescence, i.e., the photon counts originating from specific cellular structures, is registered in different channels depending on the wavelength. The absolute fluorescence intensity is ideally proportional[5] to the amount of introduced dye, which, in turn, scales with the concentration of the labeled structures or the internalized fluorescent nanomaterial.

9.1.1 Visualizing the Cell

Fluorescent molecules can be specifically introduced into cells by selecting one method out of a great number of various established ones. Thereby, live-cell imaging requires different approaches in contrast to the observation of fixed (preserved) cells. For live-cell imaging, cells can be transfected (i.e., modifying the genetic information) to induce transient or stable expression of fluorescent proteins linked to target structures [61]. As another approach, several fluorescently labeled compounds are commercially available, which can penetrate the outer cellular membrane and can either bind selectively to cellular organelles, or are enriched within intracellular environments being characterized by a low pH (e.g., lysosomes) or enhanced membrane potential (e.g., mitochondria). Immunofluorescence describes the usage of fluorescently labeled antibodies to identify certain antigens in a very specific manner [62]. As antibodies cannot penetrate the outer cellular plasma membrane due to their large size (around 160 kDa), only antigens which are presented on the outer cellular plasma membrane are detectable in live-cell imaging. Nonetheless, for fixed tissue, immunofluorescence is a widely used method, as cellular plasma membranes can be permeabilized by detergents, which facilitate the use of antibodies [63].

9.1.2 Nanomaterials

The interaction of nanomaterials with cells can either be measured directly (e.g., by tracing materials with fluorescent markers) or indirectly by studying cellular responses upon exposure. In image cytometry, both approaches can be combined. Relative uptake rates can be determined, nanoparticle transport can be examined by correlating their fluorescence patterns spatially with the intracellular distributions of specific cellular structures (direct approaches), and

[5] Depending on the optical properties of the fluorescent complex and the instrumentation.

in addition, changes in cellular morphology and functioning can be investigated (indirect approach).

9.2 Image Acquisition and Image Resolution

The value of the data which are obtained by image cytometry is strongly dependent on the capabilities of the optical system used for imaging. For meaningful interpretations of the results, one has to be aware of the capabilities and limits of the image acquisition system. Unfortunately, a perfect visual copy of the fluorophore distribution inside a specimen cannot be obtained. Every image acquired with an optical system without super-resolution capabilities is blurred due to the system's characteristic point spread function (PSF). The PSF describes how a single point source is seen by the detector in any optical system, influenced by the diffraction-limited nature of photon propagation.

Due to the relatively large spatial dimension of the PSF regarding a widefield fluorescence microscope, images acquired from any fluorophore in the specimen are blurred, because many undesired photons from unfocussed optical sections are included. Hence, the detection volume in such a system can hardly be quantified.

This problem can be circumvented by acquiring several optical slices around the desired axial position. Subsequently, the blur is reassessed to its origin (location of the fluorophore) to inverse the effects of the PSF by means of numerical deconvolution of the image stack.

In a confocal laser scanning microscope (cLSM), the light not originating from the focus is suppressed, in contrast to a conventional widefield microscope. Firstly, due to higher detection sensitivity (use of photomultipliers instead of CCD cameras), fluorophore excitation outside of the focus is minimized by decreasing the illuminating light intensity. Secondly, photons which do not originate from the axial position defined through the focal plane are depleted by a small pinhole within the emission light path. Thereby, only the central part of every fluorophore's PSF is "cropped" and additional axial resolution is gained.

Due to their small size, classical optical imaging of nanomaterials is strongly limited by diffraction. In widefield or confocal laser scanning microscopy, the integrated fluorescent intensity originating from a certain volume can be used to calculate intracellular concentrations, although distinct nanostructures might not be resolvable when lying adjacent to it. The fluorescence readout of nanomaterials equipped with sensing capabilities can often be used to characterize their intracellular environment. This often correlates with their intercellular location, although imaging is limited by diffraction [64].

9.3 Image Processing

For image segmentation (Image Segmentation, Sect. 9.4), uniform datasets are required. Therefore, appropriate handling of artifacts originating either from the optical imaging itself or from the digitizing of the underlying signals is needed to minimize intensity nonuniformities. Possible error sources have to be identified and considered during image restoration. In case of confocal fluorescence laser scanning microscopy, images are only slightly blurred by out-of-focus information, but suffer from nonuniform illumination and noise [65]. In the latter case, especially Poisson-distributed shot noise originating from photon detection at low count rates is unavoidable. Examples of methods correcting for nonuniform illumination are (i) the morphological opening of the corresponding image for background extraction, (ii) the subtraction of a blurred version of the image from the original one, or (iii) the adaption of a parameterized surface or grid of cubic splines[6] to the image and normalization of the intensity values based on the computed fit (Fig. 9.1) [66, 67]. All methods have advantages and disadvantages. Especially the first approach requires knowledge about the size distribution of the structures to be segmented. Image restoration regarding shot noise is typically performed by

[6] Splines are piecewise-defined polynomial functions.

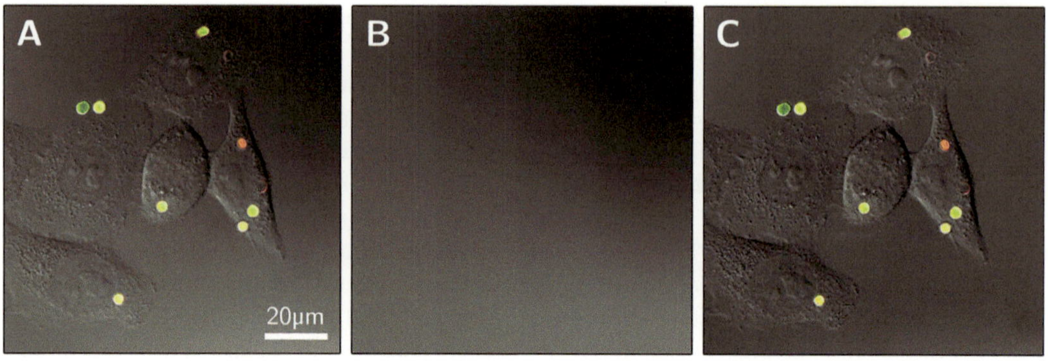

Fig. 9.1 Correction of image nonuniformities due to misaligned illumination. (**a**) Fluorescence microscope image showing human cervical cancer cells (HeLa) cells with internalized microcapsules (*green, red*) and nonuniform background in the transmission channel. (**b**) Spline-surface fit to the background. (**c**) Corresponding image after background subtraction

deconvolution or filtering [66, 68]. Noise reduction by deconvolution typically yields better results. While working with large image sets, this approach requires excessive computational time, and hence, Gaussian smoothing or especially median filtering is often favored.

9.4 Image Segmentation

By means of image segmentation, a digital image is partitioned into its constituent regions to locate objects or certain patterns. Starting from early age, the visual cortex in our brain is trained to identify and allocate objects in the image stream generated by our visual system. Although the human brain can easily recognize the boundaries of an individual cell inside tissue under the microscope, segmentation remains the most difficult task in computer vision [66]. For each segmentation problem, the image constituents are modeled (e.g., stained nuclei are bright and round). Based on this model, the segmentation algorithms are selected. In the following paragraphs, several segmentation methods most often used are briefly introduced. For practical application in image cytometry, a combination of several segmentation methods is typically used in combination with morphological image processing based on the theory of mathematical morphology. The latter case comprises the application of nonlinear operations which alter shape (shrinking/expanding) or morphology (hole filling, gap closing, intersectioning) of features in an image [66].

9.4.1 Thresholding

In the simplest case, image structures of interest (for instance, particles or cell nuclei) are well-separated and brighter than the background. Segmentation is performed by finding all connected components brighter than a suitable threshold (Fig. 9.2). Uniform image datasets are favored where all images were acquired under exactly the same conditions, and one global and manually set threshold can be used to segment all structures of interest. For more complex problems, several approaches exist in literature to determine appropriate thresholds locally [69]. Clumped objects are not separated by thresholding.

9.4.2 Watershed Segmentation and Voronoi-Based Approaches

Confluent cells, for instance, are clustered and can barely be divided and segmented by thresholding (Thresholding, Sect. 9.4.1). For such complex structures, watershed segmentation [70, 71] or Voronoi-based segmentation [72] has been proven to be very useful. Depending on the staining,

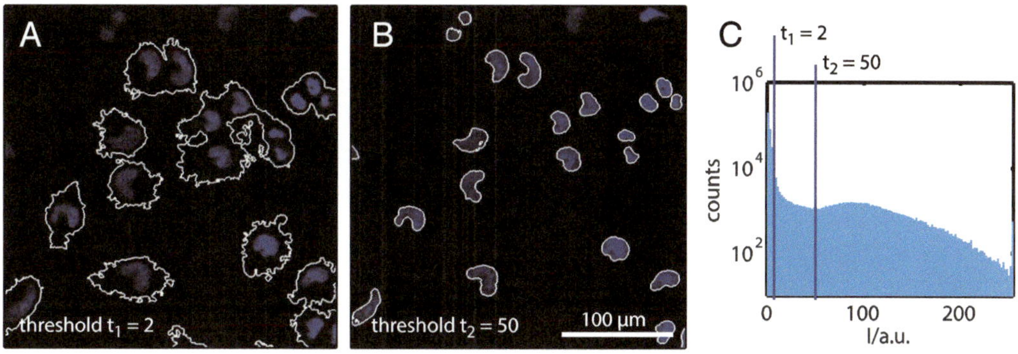

Fig. 9.2 Segmentation by thresholding. (**a**, **b**) Different thresholds were applied to a fluorescence image showing cell nuclei. (**c**) Histogram in logarithmic scale of the fluorescence image shown in (**a**) and (**b**) with the corresponding thresholds

cells are typically (i) less intense at the borders in comparison to the average intensity at the perinuclear region, or the opposite is true where (ii) cell outlines show a strong contrast and bright intensity. The first case is obtained when staining the cytoplasm, whereas a more inhomogeneous pattern is typically achieved after application of cytoskeletal stains. In the latter case, especially ruffles along the outer membrane are highlighted.

"Watershedding" requires a gradient intensity toward the object borders. Thereby, when image intensity is interpreted as topographic relief, cells can be thought of as mountains separated by valleys in such an intensity landscape. Watershed segmentation can be imagined as submerging the "image landscape" in water, i.e., filling all local minima, and creating boundaries along the lines where different water sources meet in case the water gauge is increased locally and different catchment basins are going to be connected [70, 71]. Direct application of the watershed algorithm, as sketched above, leads to oversegmentation (i.e., detection of an erroneous high number of separated regions) due to noise and local gradient irregularities [66]. In digital image cytometry, this problem is normally solved by providing the algorithm with "seeds" based on the coordinates of unique cellular structures from a parallel image. In case nuclei are stained, they serve as superior markers ("primary objects"), usually being well-separated and easily segmentable by applying a global threshold (Fig. 9.3d).

Voronoi-based segmentation also requires a set of primary objects limiting the number and constraining the position of potential "secondary objects." For each seed, a discretized approximation of its Voronoi region[7] (Fig. 9.3e, f) is calculated on a manifold with a metric controlled by local image features [72].

9.4.3 Shape-Based Segmentation

For segmentation of objects which are either not separated by less intense borders or when no markers are provided, the inclusion of additional features into the segmentation model is required in order to transform the image structures into other ones, which can be segmented by simple peak-finding algorithms.

The Hough transform is a feature extraction technique, which can be used to emphasize structures of any shape [66, 74]. In the case of analytically describable shapes, such as lines or circles, a weight is assigned to each pixel of an image, which can be seen as the "probability" of being the origin of an earlier defined parameterized pattern.

For the detection of circular structures (Circle Hough Transform, CHT), for example, the sum of pixel intensities along a circle of radius r

[7]Voronoi diagrams describe a distance-controlled partitioning of a plane into regions based on seeds, cf. Figure 9.3e [73].

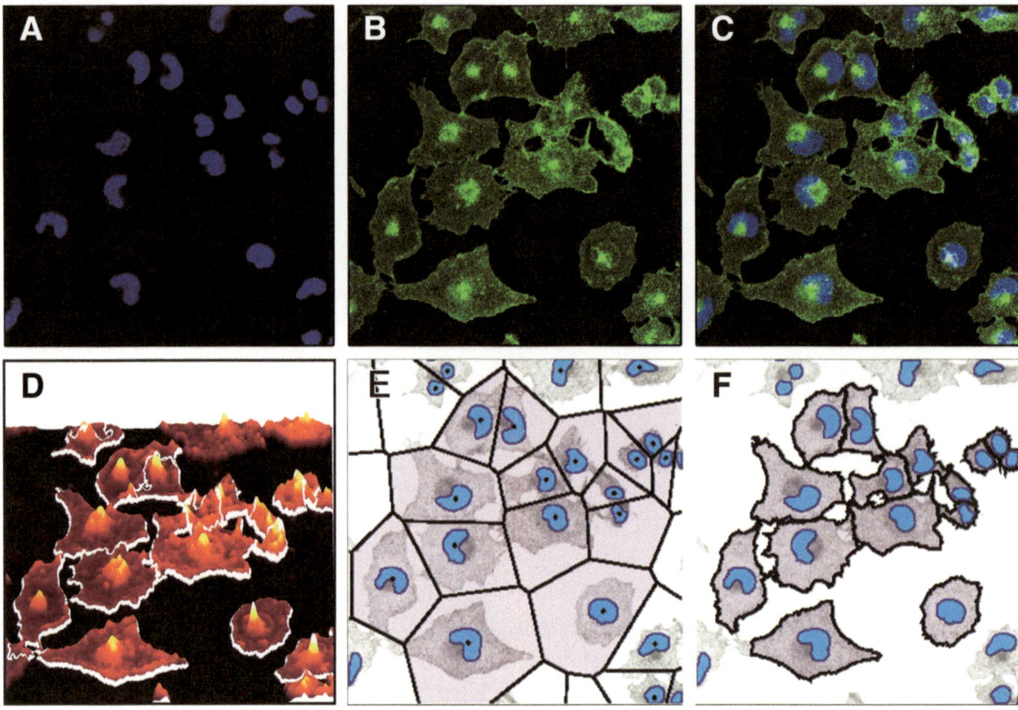

Fig. 9.3 Segmentation of cells. (**a–c**) Two channels of a fluorescence image of fixed HeLa cells stained with Hoechst 33342 (nuclei, *blue* channel, (**a**) and with fluorescently labeled wheat germ agglutinin (plasma membrane, *green* channel, (**b**). The overlay of both channels is shown in (**c**). (**d**) The result of seeded (seeds were obtained from the coordinates of the nuclei shown in (**a**) watershed segmentation on a Sobel-filtered (edge enhanced) version of the image shown in (**b**). (**e**) Voronoi diagram [73] based on the positions of the nuclei. (**f**) Voronoi-based segmentation as described by Jones et al. [72]. For comparison, corresponding objects in (**e**) and (**f**) are shaded and objects touching the border are not considered

around each pixel px_i is calculated for each pixel, yielding the two-dimensional so-called accumulator matrix. In this representation, pixels, which are the origins of circular structures of radius r in the original image, appear as bright spots (Fig. 9.4). By finding the coordinates of the local maxima in the accumulator matrix, circular structures are registered. In most cases the last task requires additional post-processing and filtering to suppress unwanted side lobs. The CHT is extremely helpful to segment spherical particles in microscopic images, which neither show a peak with Gaussian intensity distribution nor occur in clusters nor are aggregated (Fig. 9.4). By extending the CHT algorithm, the identification of circular objects bearing different sizes is possible (e.g., fluorescently labeled polymer capsules as demonstrated in [29], cf. Fig. 9.5).

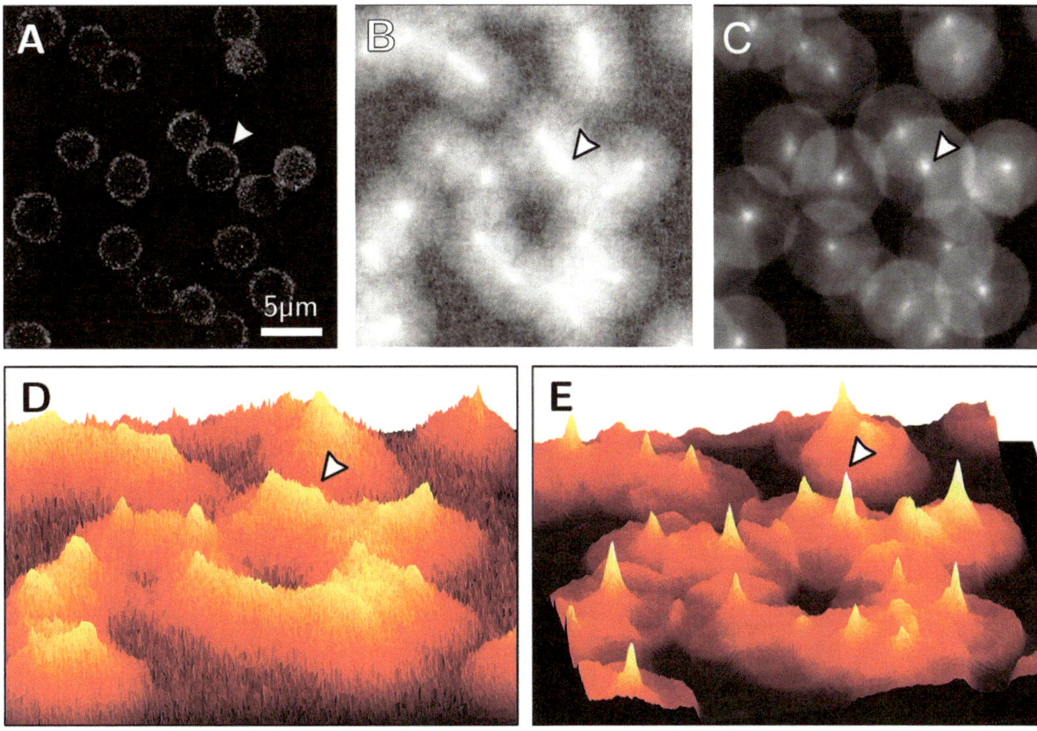

Fig. 9.4 Circle Hough Transform. (**a**) Noisy fluorescence image of hollow and aggregated microcapsules. (**b**, **d**) Accumulator matrix return from a classical Circle Hough Transform for circles with d = 4.2 μm. (**c**, **e**) Accumulator matrix obtained from a modified algorithm (Fig. 9.5) for identification of center coordinates for capsules with d <7 μm. For registration of the different images, one capsule is highlighted with an *arrow*

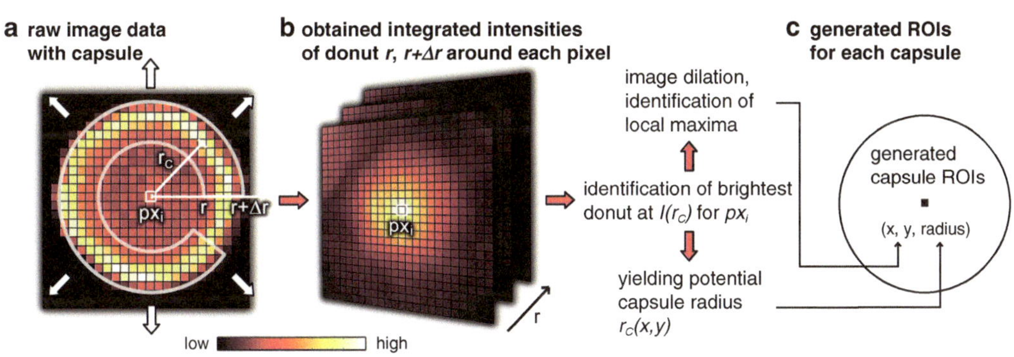

Fig. 9.5 Diameter-detecting, modified Circular Hough Transform. (**a**) In fluorescence micrographs, hollow microcapsules appear as circular objects with increased intensity along the shell. By determining the integrated intensity I along a donut or radius r and thickness Δr for each px_i, the function I(r, Δr) is obtained. When "finding" a shell with origin at px_i and radius r_C, $I(r_C)$ is strongly increased. (**b**) $I(r_C)$ is assigned to the accumulator matrix (Fig. 9.4e). (**c**) Coordinates and radius r_C of the detected structure are obtained

9.5 Feature Extraction and Measurements

Several descriptive features can be extracted from segmented, individual objects in microscopic images. An overview is given by Rodenacker et al. [75]. Features are either based on the spatial pixel arrangement and describe the shape (morphometric features, cf. Table 9.1), give information about the distribution of pixel intensities (densiometric features, cf. Fig. 9.6), or describe the spatial variations of pixel intensities (textural or structural features, cf. Table 9.2). If the microscope images comprise several spectral channels as in the case of fluorescence microscopy, segmented cell objects based on nuclei and cytoplasm (Watershed Segmentation and Voronoi-Based Approaches, Sect. 9.4.2) can be used to calculate densiometric or textural properties at another spectral region. In other words, the identified objects are used to mask the information in other image channels. By doing so, the spatial intracellular arrangement of tertiary structures can be obtained. Also the level of certain dyes can be observed and related to protein concentrations or expression levels, or the uptake rate of nanomaterials can be quantified. All properties can be related back to the underlying object (i.e., cell) and tracked over time (Object Tracking and Digital Video Analysis, Sect. 9.7) in case of live-cell imaging [57, 76–78].

Table 9.1 Examples for morphometric features

Feature	Perimeter $d = 1\mu m$ 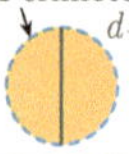	$0.5\mu m$ 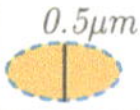	$0.25\mu m$ 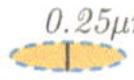	$0\mu m$
$A/\mu m^2$	0.78	0.39	0.2	0
$P/\mu m$	3.1	2.5	2.2	2
F	1	0.76	0.5	0
Z_0	1	0.5	0.25	0

A area, *P* perimeter, *F* form factor = $4\pi A/P^2$, Z_0 Zernike moment of 0th order. Zernike moments describe the decomposition of an image object onto an orthogonal set of polynomials similar to the way that Fourier coefficients are used to decompose a time series [60]. Similar to the form factor F, the 0th moment Z_0 can be used to describe whether a shape is similar to a disk ($Z_0 = 1$) or more spindle like ($Z_0 = 0$). *d* corresponds to the semiminor axis of the example shapes, if being represented by an ellipse

9.6 Feature Correlation

Different approaches exist to investigate the spatial arrangement of intracellular structures from fluorescence microscope images. With the measures introduced below, the degree of colocalization of different patterns being captured in two different fluorescence channels can be quantified (Table 9.3) [81].

9.6.1 Intensity-Based Correlation

Pearson's correlation coefficient R_r can be used to determine the similarity of two patterns. In the context of digital image cytometry, R_r (Eq. 8.1) can be calculated based on the patterns visible in two distinct fluorescence channels either per image or per underlying cell object (in case statements regarding different cell populations are needed). Pearson's correlation coefficient is defined as the covariance of the intensity values of the two patterns divided by the product of their standard deviations and is widely used in pattern recognition [66].

Table 9.2 Examples for textural features. Textural features can be used to describe the fine-structure of actin and tubulin staining of cells

Texture	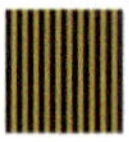				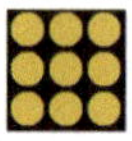
T_{cont}/a.u.	36	4.2	0.6	0	7.8
T_{corr}/a.u.	−0.5	−0.6	0.3	0	0.7

T_{cont} texture contrast, T_{corr} texture correlation referring to Haralick et al. [79]

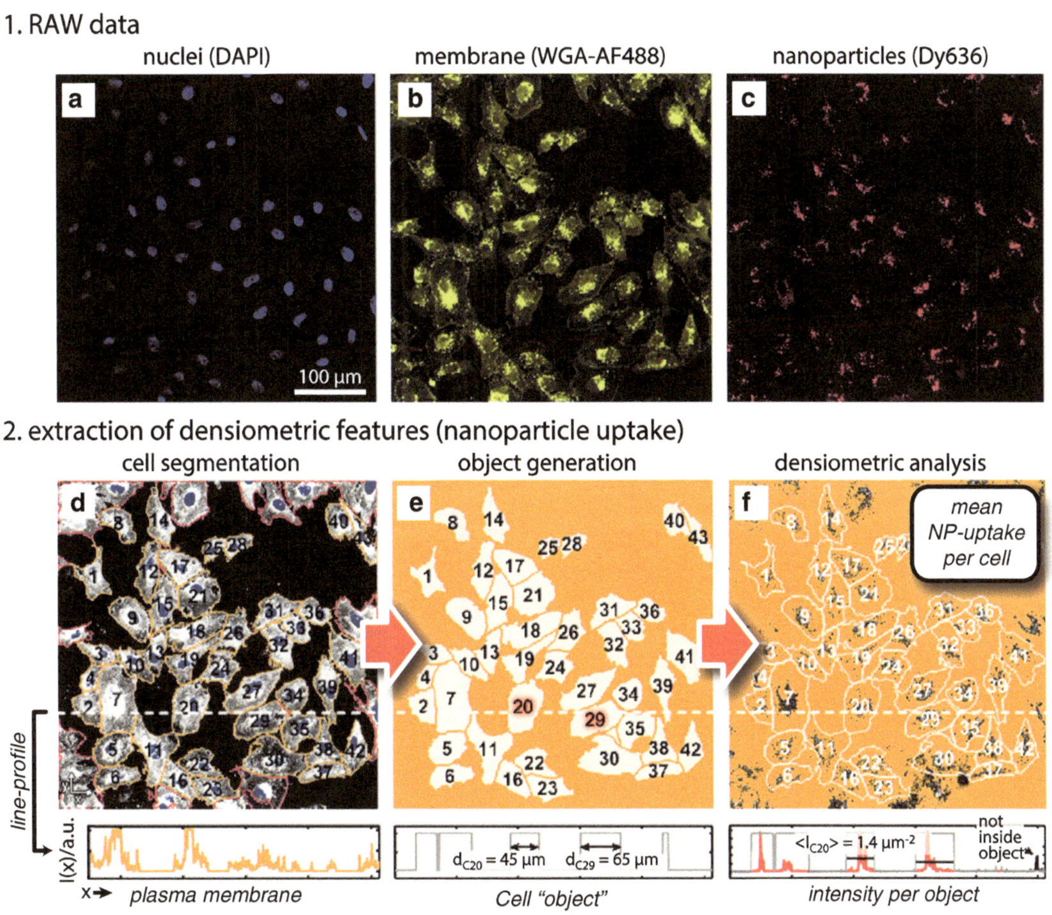

Fig. 9.6 Example for densiometric features extraction. Illustration of the image processing steps to obtain the densiometric feature "integrated intensity" of nanoparticles associated with cells (objects) imaged in an additional fluorescence channel. (**a**, **b**) For Voronoi-based cell segmentation, images of the nuclei (stained with 4′,6-diamidino-2-phenylindole, DAPI, *blue*) and of the outer plasma membrane (stained with AlexaFluor488-labeled wheat germ agglutinin, WGA-AF488, *yellow*) were used. (**c**) Associated red fluorescence signal of internalized nanoparticles. (**d**) Results of the segmentation procedure. At the bottom, the line intensity profile of the plasma membrane stain along the *dashed line* is shown. (**e**) Shapes of the obtained cell objects. At the bottom the line intensity profile of the cell objects along the *dashed line* is shown, and cell #20 and #29 are highlighted. (**f**) Overlay of cell object outlines and nanoparticle signal. The integrated nanoparticle intensity I_{Int} is calculated per cell (densiometric feature) and assigned to the corresponding object. Accordingly, nanoparticles outside cell objects are not considered. In general, the integrated intensity is proportional to the total amount of a fluorescent compound per object. Thus, in this case, the integrated nanoparticle intensity I_{Int} can be related to the total uptake of nanoparticles. For each cell object, I_{Int} is calculated as the mean NP intensity $\langle I \rangle$ per cell × the area of each object. For clarification in 1D, the total uptake I_{Int} along the line profile would be determined as the object length d times the mean NP intensity $\langle I \rangle$ within the corresponding object (example calculation for cell #20: $I_{Int,C20} = \langle I_{C20} \rangle \cdot d_{C20}$ = 47.6 NP intensity units/µm^{-1}) (Reprinted (adapted) with permission from Pelaz et al.[80]. Copyright (2015) American Chemical Society)

$$R_r = \frac{\Sigma\,(R_i - \bar{R})\cdot(G_i - \bar{G})}{\sqrt{\Sigma\,(R_i - \bar{R})^2 \cdot \Sigma\,(G_i - \bar{G})^2}} \in [-1,1] \qquad (8.1)$$

Considering two fluorescent channels R and G, then R_i or G_i, respectively, is the intensity of the ith voxel, while $\bar{R}$ and $\bar{G}$ are the mean values of all voxel intensities in the corresponding channel. A positive value for R_r indicates a high degree of colocalization or high pattern similarity, while negative values indicate exclusion. As the average image intensities are included, this coefficient is only slightly biased by different background levels of the two images [81]. If the correction for the average image intensities is not performed (e.g., to compare different labeling efficiencies), then Manders' colocalization coefficient M (Eq. 8.2) is obtained [82].

$$M = \frac{\sum (R_i \cdot G_i)}{\sqrt{\sum R_i^2 \cdot \sum G_i^2}} \in [0,1] \qquad (8.2)$$

9.6.2 Object-Based Correlation

In object-based correlation, the spatial arrangement of objects in two distinct channels is analyzed. Therefore, firstly, both images need to be binarized by an appropriate segmentation routine (e.g., thresholding) before calculation of either R_r or M. Still, the underlying intensity values of the objects can be used as weightings.

In cases of asymmetrical colocalization (Table 9.3, Example 4) where Pearson's or Manders' coefficients are less meaningful, the use of Manders' distinct overlap coefficients M_1 and M_2 (Eq. 8.3) might make more sense to quantify the spatial overlap of two patterns [82]. Segmentation is needed to decide whether a voxel is colocalizing or not.

$$M_1 = \frac{\Sigma\, R_{i,coloc}}{\Sigma\, R_i} \in [0,1] \text{ and } M_2 = \frac{\Sigma\, G_{i,coloc}}{\Sigma\, G_i} \in [0,1] \qquad (8.3)$$

Only pixel intensities $R_{i,coloc}$ of pixels colocalizing with an object in the opposite channel are considered. R_i or G_i, respectively, is the intensity of the ith voxel in the corresponding channel.

Table 9.3 Exemplarily calculated correlation coefficients for representative patterns

Example	Type		R_r	M	M_1	M_2	$\bar{I}_G(O_R)$	$\bar{I}_G(O_R)$
1	Separated	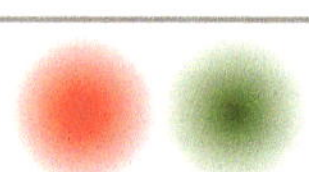	−0.34	0	0	0	0	0
2	Partial overlap		−0.03	0.2	0.42	0.42	35.5	35.5
3	Overlap	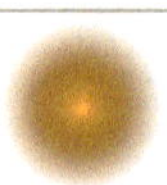	1	1	1	1	85.1	85.1
4	Inclusion	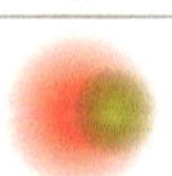	0.46	0.52	0.42	1	31.1	36.7

R_r Pearson's correlations coefficient [66], M Manders' colocalization coefficient, [82] M_1 and M_2 Manders' distinct overlap coefficients [82], and $\bar{I}_R(O_G)$ and $\bar{I}_G(O_R)$ for quantification of the average pixel intensity along objects in channel R or G, respectively [36]. The bit-depth of the example images was 8 resulting in a maximum intensity value of 255. All patterns exhibited a linear gradient from $I_{max} = 255$ to $I_{min} = 0$. The segmentation method to determine colocalizing pixels in case of the determination of M_1, M_2, and $\bar{I}_R$ and $\bar{I}_G$ was based on thresholding with a threshold of 1

By dividing the sum of intensities from all colocalizing voxels ($R_{i,coloc}$ or $G_{i,coloc}$, based on Eq. 8.3) by the number N of colocalizing voxels instead of by the sum of all pixel intensities of the corresponding channel, the average fluorescent intensity $\bar{I}$ along all objects O in the opposite channel (O_G or O_R) can be calculated (Eq. 8.4).

$$\bar{I}_R\left(O_G\right)=\frac{\sum R_{i,coloc}}{N_{coloc}}\in\mathbb{R}_+ \text{ and } \bar{I}_G\left(O_R\right)=\frac{\sum G_{i,coloc}}{N_{coloc}}\in\mathbb{R}_+ \quad (8.4)$$

In case of quantifying cell uptake rates of nanomaterials, the last equations (Eq. 8.4) are rather useful to assess the density of fluorescent nanomaterials measured, for instance, in channel R along the certain objects (e.g., fluorescence-labeled lysosomes) imaged in channel G, i.e., $\bar{I}_R(O_G)$, respectively.

9.7 Object Tracking and Digital Video Analysis

Trajectories of individual objects can be extracted from time-lapse fluorescence micrographs by digital video analysis [83]. The time evolution of the distribution of objects (Eq. 8.5) can be used to determine the progression of certain features associated with the objects over time on the level of individual objects (e.g., cells or particles).

$$\rho\left(\vec{r},t\right)=\sum_{N}^{i=1}\delta\left(\vec{r}-\vec{r}_i\left(t\right)\right) \quad (8.5)$$

In Eq. 8.5, $\vec{r}_i\left(t\right)$ represents the location of the ith object in a field of N particles at time t. In each frame in a sequence of video images, the objects' coordinates and corresponding features (Feature Extraction, Sect. 9.5) are identified by segmentation (Image Segmentation, Sect. 9.4). The trajectories $\rho\left(\vec{r},t\right)$ are produced by matching up locations in each image with corresponding locations in latter images. To link objects in two successive frames, the most probable set of N identifications between N locations in two consecutive images is required. Models of the underlying dynamics (e.g., Brownian motion for particles) are often considered to increase corrected linking of object coordinates. In addition, unique object features might be included into the probability calculations. Finally, gap closing, merging, and splitting steps are needed to correctly handle objects missing in certain video frames (i.e., out of focus) [78, 83, 84].

9.8 Conclusion

Digital image cytometry can be a powerful tool which simplifies the assessment of processes on the cellular and subcellular level based on high-throughput fluorescence microscopy and image processing. It is closely related to flow cytometry, but in comparison to these techniques, the list of accessible cell features is increased dramatically [41, 43]. The cell segmentation in flow cytometry is "solved" by subsequent passing of individual cells through the exciting laser beam. Accordingly, cell recognition in digital image cytometry is more challenging and requires specific stainings in combination with sophisticated computer vision algorithms. Inappropriate segmentation parameters may lead to inaccurate results including artifacts and/or methodical errors.

In addition the endpoints of the assays have to be selected carefully. The classical mistake which can be made (also in classical flow cytometry) is caused by cytotoxicity-induced cell loss. The profile of the remaining cells does not represent the original population, as the residual cells might behave abnormally in some way making them resistant to the toxic impulse.

The major advantage of digital image cytometry in comparison to flow cytometric approaches is the ability to "look into the cell" in high spatial resolution, to examine cells in their natural state,[8] and to measure kinetics. After the measurement, an individual cell is not lost and can be examined again at a later point in time. This can be used either (i) to determine the evolution of global features, i.e., similar to measuring several samples

[8] For instance, in the case of adherent cells, no detachment and transfer into certain buffers prior to cytometric measurements are required.

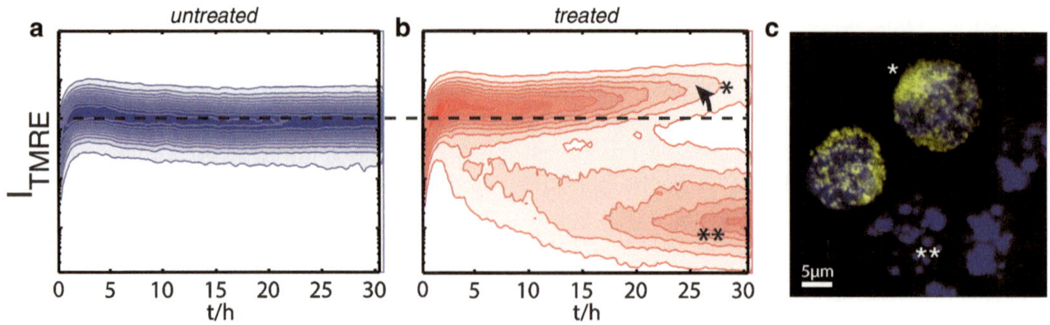

Fig. 9.7 Digital image cytometry for time-resolved densiometric measurements. The mitochondrial membrane potential $\Delta\psi_m$ of human promyelocytic leukemia cells (HL-60) upon treatment with a chemotherapeutic agent cytarabine (AraC) is indicated by the fluorescence of the dye tetramethylrhodamine ethyl (I_{TMRE}). TMRE and AraC were added at $t=0$ min. (**a**) In untreated control cells, the mitochondrial membrane potential is not affected. (**b**) In treated cells hyperpolarization of mitochondrial membranes can be observed before apoptosis occur. The part of the intensity distribution representing cells with hyperpolarized mitochondrial membranes is marked with (*); the part representing apoptotic cells is labeled with (**). The *dashed line* is drawn to allow comparison of the I_{TMRE} values between treated and untreated cells. (**c**) Fluorescence micrograph showing cells in suspension with high membrane potential (*yellow*, *) and apoptotic cells with depolarized mitochondrial membranes (**). Nuclei were stained in *blue* (Hoechst 33342). In this Figure unpublished data are shown for the purpose of illustration

representing different points in time with the flow cytometer, or (ii) for tracking of individual cells and evaluation of certain features on the single cell level over time. An example for the first option is shown in Fig. 9.7 where the mitochondrial membrane potential (reported by a fluorescence dye) upon treatment with a chemotherapeutic agent is assessed in human promyelocytic leukemia cells (HL-60) time-dependently. From the data the evolution of different cell populations (cells with hyperpolarized and depolarized mitochondrial membranes) can be observed in a high temporal resolution. Every outlier can be traced back to the underlying image and finally to the underlying cell object.

Still, for all these kinds of measurements, the segmentation of cells in every single image frame is required. This implies that on the one hand, the staining techniques have to be optimized carefully to avoid any interference with the cell viability and the actual measurements. On the other hand, large quantities of multidimensional image data whose processing is time-consuming and requires computing power are produced for automatic segmentation and feature extraction. Finally, data evaluation and an appropriate representation of the obtained results are a challenge, as the datasets are highly multidimensional.

For segmentation of the image data acquired from living cells, DNA stains (e.g., Hoechst 33342), commonly used for identification of primary cell nuclei (Image Segmentation, Sect. 9.4), can cause problems, since they interfere with DNA replication and exhibit phototoxicity [85]. Similar problems can be attributed to membrane stains, as certain receptors might be blocked or undesired cellular responses might be triggered. Consequently, the stain concentrations should always be kept as low as possible even if the quality of the acquired images is reduced by low fluorescence signals. Drawbacks in image quality can usually be solved with appropriate image restoration algorithms or are of no consequence due to the high number of analyzed cells.

A very important point for the successful application of digital image cytometry is the conceptual design of the experiment. Almost all experimental and technical parameters are interrelated. For instance, the fluorescence characteristics of nanomaterials should not interfere with the dyes introduced for later cell segmentation. Image resolution is competing with temporal resolution which in turn is limited by the total cell count and the number of different conditions/samples (e.g., wells) to be captured. High cell numbers are desired for high statistical significance. For cell tracking, quite a high temporal resolution is needed for correct cell identification in consecutive time-lapse image frames. On the contrary, a high temporal resolution also limits the total cell count.

Recently, several optical "super-resolution" methods have been developed that are capable of resolving nanostructures down to several tens of nanometers [86, 87]. The concept of digital image cytometry presented aims at generating data that represents thousands of individual cells. Yet, super-resolution microscopes are rather slow and hard to automatize. In addition, when covering a comparable growth area with a similar number of cells, the data output would be extreme and slow to process with conventional work stations. Realistically, imaging is limited to subcellular structures or macromolecules in this case. Then, the challenge of image segmentation lies more in recognizing different intracellular compartments than in the detection of whole cells. However, when assessing the cellular interaction with nanomaterials, it is often not even necessary to resolve individual particles as the cellular response is well-detectable.

In a nutshell, high-throughput microscopy in combination with digital image cytometry can help to answer the following questions with high statistical relevance:

1. How many nanoparticles are internalized?
2. Where they are intracellularly transported to?
3. How do they affect cells?

Within the field of nanobiotechnology particle-cell interactions, intracellular release, sensor particle readout, and particle-induced cellular responses are generally suitable problems for future investigation aided by the introduced methodology. The development of serious nanomedicine is an emerging and fast-growing field. Hence, reliable and sensitive assays are needed to probe nanoparticle functioning and cytotoxicity at an early stage, where digital image cytometry does function as a valuable research tool.

References

1. Goesmann H, Feldmann C (2010) Nanoparticulate functional materials. Angew Chemie Int Ed 49:1362–1395
2. Jain PK, Huang X, El-Sayed IH, El-Sayed MA (2008) Noble metals on the nanoscale: optical and photothermal properties and some applications in imaging, sensing, biology, and medicine. Acc Chem Res 41:1578–1586
3. Barreto JA, O'Malley W, Kubeil M et al (2011) Nanomaterials: applications in cancer imaging and therapy. Adv Mater 23:H18–H40
4. Guo D, Xie G, Luo J (2014) Mechanical properties of nanoparticles: basics and applications. J Phys D Appl Phys 47:013001
5. Daniel MC, Astruc D (2004) Gold nanoparticles: assembly, supramolecular chemistry, quantum-size-related properties, and applications toward biology, catalysis, and nanotechnology. Chem Rev 104:293–346
6. Brannon-Peppas L, Blanchette JO (2012) Nanoparticle and targeted systems for cancer therapy. Adv Drug Deliv Rev 64:206–212
7. McCarthy JR, Kelly KA, Sun EY, Weissleder R (2007) Targeted delivery of multifunctional magnetic nanoparticles. Nanomedicine 2:153–67
8. Liong M, Lu J, Kovochich M et al (2008) Multifunctional inorganic nanoparticles for imaging, targeting, and drug delivery. ACS Nano 2:889–896
9. Jun YW, Lee JH, Cheon J (2008) Chemical design of nanoparticle probes for high-performance magnetic resonance imaging. Angew Chemie - Int Ed 47:5122–5135
10. Casula MF, Floris P, Innocenti C et al (2010) Magnetic resonance imaging contrast agents based on iron oxide superparamagnetic ferrofluids. Chem Mater 22:1739–1748
11. Dutz S, Andrä W, Hergt R et al (2007) Biomedical heating applications of magnetic iron oxide nanoparticles. In: Magjarevic R (ed) World congress on medical physics and biomedical engineering 2006 SE - 76. Springer, Heidelberg, pp 271–274
12. Sperling RA, Rivera Gil P, Zhang F et al (2008) Biological applications of gold nanoparticles. Chem Soc Rev 37:1896–1908
13. Xia Y (2008) Nanomaterials at work in biomedical research. Nat Mater 7:758–760
14. Somers RC, Bawendi MG, Nocera DG (2007) CdSe nanocrystal based chem-/bio- sensors. Chem Soc Rev 36:579–591
15. Klostranec JM, Chan WCW (2006) Quantum dots in biological and biomedical research: recent progress and present challenges. Adv Mater 18:1953–1964
16. Murphy CJ, Gole AM, Stone JW et al (2008) Gold nanoparticles in biology: beyond toxicity to cellular imaging. Acc Chem Res 41:1721–1730

17. Saha K, Agasti SS, Kim C et al (2012) Gold nanoparticles in chemical and biological sensing. Chem Rev 112:2739–2779
18. Lévy R, Shaheen U, Cesbron Y, Sée V (2010) Gold nanoparticles delivery in mammalian live cells: a critical review. Nano Rev 1
19. Thiesen B, Jordan A (2008) Clinical applications of magnetic nanoparticles for hyperthermia. Int J Hyperthermia 24:467–474
20. Jain PK, ElSayed IH, El-Sayed MA (2007) Au nanoparticles target cancer. Nano Today 2:18–29
21. Neely A, Perry C, Varisli B et al (2009) Ultrasensitive and highly selective detection of alzheimer's disease biomarker using two-photon rayleigh scattering properties of gold nanoparticle. ACS Nano 3:2834–2840
22. Fan J, Li H, Jiang J et al (2008) 3c-sic nanocrystals as fluorescent biological labels. Small 4:1058–1062
23. Zhang S, Li J, Lykotrafitis G et al (2009) Size-dependent endocytosis of nanoparticles. Adv Mater 21:419–424
24. Stoehr LC, Gonzalez E, Stampfl A et al (2011) Shape matters: effects of silver nanospheres and wires on human alveolar epithelial cells. Part Fibre Toxicol 8:36
25. Hühn D, Kantner K, Geidel C et al (2013) Polymer-coated nanoparticles interacting with proteins and cells: focusing on the sign of the net charge. ACS Nano 7:3253–3263
26. Harush-Frenkel O, Rozentur E, Benita S, Altschuler Y (2008) Surface charge of nanoparticles determines their endocytic and transcytotic pathway in polarized mdck cells. Biomacromolecules 9:435–443
27. Schweiger C, Hartmann R, Zhang F et al (2012) Quantification of the internalization patterns of superparamagnetic iron oxide nanoparticles with opposite charge. J Nanobiotechnology 10:28
28. Anguissola S, Garry D, Salvati A et al (2014) High content analysis provides mechanistic insights on the pathways of toxicity induced by amine-modified polystyrene nanoparticles. PLoS One 9:e108025
29. Hartmann R, Weidenbach M, Neubauer M et al (2015) Stiffness-dependent in vitro uptake and lysosomal acidification of colloidal particles. Angew Chemie Int Ed 54:1365–1368
30. Banquy X, Suarez F, Argaw A et al (2009) Effect of mechanical properties of hydrogel nanoparticles on macrophage cell uptake. Soft Matter 5:3984
31. Liu W, Zhou X, Mao Z et al (2012) Uptake of hydrogel particles with different stiffness and its influence on hepg2 cell functions. Soft Matter 8:9235
32. Cedervall T, Lynch I, Lindman S et al (2007) Understanding the nanoparticle-protein corona using methods to quantify exchange rates and affinities of proteins for nanoparticles. Proc Natl Acad Sci U S A 104:2050–2055
33. Nel AE, Mädler L, Velegol D et al (2009) Understanding biophysicochemical interactions at the nano-bio interface. Nat Mater 8:543–557
34. Monopoli MP, Aberg C, Salvati A, Dawson KA (2012) Biomolecular coronas provide the biological identity of nanosized materials. Nat Nanotechnol 7:779–786
35. Jiang X, Röcker C, Hafner M et al (2010) Endo- and exocytosis of zwitterionic quantum dot nanoparticles by live hela cells. ACS Nano 4:6787–6797
36. Ma X, Hartmann R, Jimenez de Aberasturi D, Yang F, Soenen S. J, Manshian B, Franz J, Valdeperez D, Pelaz B, Hampp N et al. (n.d.) Nano Today, in revision.
37. Kim JA, Åberg C, Salvati A, Dawson KA (2011) Role of cell cycle on the cellular uptake and dilution of nanoparticles in a cell population. Nat Nanotechnol 7:62–68
38. De Jong WH, Hagens WI, Krystek P et al (2008) Particle size-dependent organ distribution of gold nanoparticles after intravenous administration. Biomaterials 29:1912–1919
39. Van Der Zande M, Vandebriel RJ, Van Doren E et al (2012) Distribution, elimination, and toxicity of silver nanoparticles and silver ions in rats after 28-day oral exposure. ACS Nano 6:7427–7442
40. Kreyling WG, Abdelmonem AM, Ali Z et al (2015) In vivo integrity of polymer-coated gold nanoparticles. Nat Nanotechnol 10:619–623
41. Haney SA (2007) High content screening: science, techniques and applications. High Content Screen Sci Tech Appl 1:391
42. Gasparri F (2009) An overview of cell phenotypes in hcs: limitations and advantages. Expert Opin Drug Discov 4:643–657
43. Taylor DL, Haskins JR, Giuliano KA (2007) High content screening : a powerful approach to systems cell biology and drug discovery
44. Richards GR, Smith AJ, Parry F et al (2006) A morphology- and kinetics-based cascade for human neural cell high content screening. Assay Drug Dev Technol 4:143–152
45. Fennell M, Chan H, Wood A (2006) Multiparameter measurement of caspase 3 activation and apoptotic cell death in nt2 neuronal precursor cells using high-content analysis. J Biomol Screen Off J Soc Biomol Screen 11:296–302
46. Inglefield JR, Larson CJ, Gibson SJ et al (2006) Apoptotic responses in squamous carcinoma and epithelial cells to small-molecule toll-like receptor agonists evaluated with automated cytometry. J Biomol Screen Off J Soc Biomol Screen 11:575–585
47. Lövborg H, Gullbo J, Larsson R (2005) Screening for apoptosis--classical and emerging techniques. Anticancer Drugs 16:593–599
48. Bertelsen M, Sanfridson A (2005) Inflammatory pathway analysis using a high content screening platform. Assay Drug Dev Technol 3:261–271
49. Lang P, Yeow K, Nichols A, Scheer A (2006) Cellular imaging in drug discovery. Nat Rev Drug Discov 5:343–356
50. Moffat J, Grueneberg DA, Yang X et al (2006) A lentiviral rnai library for human and mouse genes applied to an arrayed viral high-content screen. Cell 124:1283–1298

51. Bjorklund M, Taipale M, Varjosalo M et al (2006) Identification of pathways regulating cell size and cell-cycle progression by rnai. Nature 439:1009–1013
52. Jan E, Byrne SJ, Cuddihy M et al (2008) High-content screening as a universal tool for fingerprinting of cytotoxicity of nanoparticles. ACS Nano 2:928–38
53. Soenen SJ, Manshian B, Montenegro JM et al (2012) Cytotoxic effects of gold nanoparticles: a multiparametric study. ACS Nano 6:5767–5783
54. Solmesky LJ, Shuman M, Goldsmith M et al (2011) Assessing cellular toxicities in fibroblasts upon exposure to lipid-based nanoparticles: a high content analysis approach. Nanotechnology 22:494016
55. Verma A, Stellacci F (2010) Effect of surface properties on nanoparticle-cell interactions. Small 6:12–21
56. Lindblad J (2002) Development of algorithms for digital image cytometry
57. Carpenter AE, Jones TR, Lamprecht MR, Clarke C, Kang IH, Friman O, Guertin DA, Chang JH, Lindquist RA, Moffat J et al (2006) Cellprofiler: image analysis software for identifying and quantifying cell phenotypes. Genome Biol 7:R100
58. Basiji DA, Ortyn WE, Liang L et al (2007) Cellular image analysis and imaging by flow cytometry. Clin Lab Med 27:653–670
59. Dreinhöfer KE, Baldetorp B, Åkerman M et al (2002) DNA ploidy in soft tissue sarcoma: comparison of flow and image cytometry with clinical follow-up in 93 patients. Clin Cytom 50:19–24
60. Boland MV, Markey MK, Murphy RF (1998) Automated recognition of patterns characteristic of subcellular structures in fluorescence microscopy images. Cytometry 33:366–375
61. Kim TK, Eberwine JH (2010) Mammalian cell transfection: the present and the future. Anal Bioanal Chem 397:3173–3178
62. Donaldson JG (2002) Immunofluorescence staining. Curr Protoc Immunol Chapter 21, Unit 21.3.
63. Suzuki T, Matsuzaki T, Hagiwara H et al (2007) Recent advances in fluorescent labeling techniques for fluorescence microscopy. Acta Histochem Cytochem 40:131–137
64. Hartmann R, Carregal-Romero S, Parak WJ, Rivera_ Gil P (2012) Investigating nanoparticle internalization patterns by quantitative correlation analysis of microscopy imaging data. In: de la Fuente JM, Grazu V (eds) Inorganic nanoparticles vs organic nanoparticles. Elsevier Ltd, Amsterdam, p 181
65. Pawley JB, Masters BR (2008) Handbook of biological confocal microscopy, third edition. J Biomed Opt 13:029902
66. Gonzalez RC, Woods RE (2007) Digital image processing. 976
67. Lindblad J, Bengtsson E (2001) A comparison of methods for estimation of intensity non uniformities in 2d and 3d microscope images of fluorescence stained cells. In: Proceedings of 12th Scandinavian conference on image analysis, 264–271.
68. Landmann L (2002) Deconvolution improves colocalization analysis of multiple uorochromes in 3d confocal data sets more than ltering techniques. Imaging 208:134–147
69. Chen S, Leung H (2004) Chaotic spread spectrum watermarking for remote sensing images. J Electron Imaging 13:220
70. Vincent L (1993) Morphological grayscale reconstruction in image analysis: applications and efficient algorithms. IEEE Trans Image Process 2:176–201
71. Meyer F, Beucher S (1990) Morphological segmentation. J Vis Commun Image Represent 1:21–46
72. Jones TR, Carpenter A, Golland P (2005) Voronoi-based segmentation of cells on image manifolds. In: Computer vision for biomedical image applications. Springer, Heidelberg, pp 535–543
73. Aurenhammer F (1991) Voronoi diagrams - a survey of a fundamental geometric data structure. ACM Comput Surv 23:345–405
74. Ballard DH (1981) Generalizing the hough transform to detect arbitrary shapes. Pattern Recognit 13:111–122
75. Rodenacker K, Bengtsson E (2003) A feature set for cytometry on digitized microscopic images. Anal Cell Pathol 25:1–36
76. Perlman ZE, Slack MD, Feng Y et al (2004) Multidimensional drug profiling by automated microscopy. Science 306:1194–1198
77. Mitchison TJ (2005) Small-molecule screening and profiling by using automated microscopy. ChemBioChem 6:33–39
78. Jaqaman K, Loerke D, Mettlen M et al (2008) Robust single-particle tracking in live-cell time-lapse sequences. Nat Methods 5:695–702
79. Haralick RM (1979) Statistical and structural approaches to texture. Proc IEEE 67:786–804
80. Pelaz B, del Pino P, Maffre P et al (2015) Surface functionalization of nanoparticles with polyethylene glycol: effects on protein adsorption and cellular uptake. ACS Nano 9:6996–7008
81. Hartmann R, Carregal-Romero S, Parak WJ, Rivera-Gil P (2012) Investigating nanoparticle internalization patterns by quantitative correlation analysis of microscopy imaging data. Front Nanosci 4:181–196
82. Manders EMM, Verbeek FJ, Aten JA (1993) Measurement of colocalization of objects in dual-color confocal images. J Microsc 169:375–382
83. Crocker J, Crocker J, Grier D (1996) Methods of digital video microscopy for colloidal studies. J Colloid Interface Sci 179:298–310
84. Jaqaman K, Danuser G (2009) Computational image analysis of cellular dynamics: a case study based on particle tracking. Cold Spring Harb. Protoc. 4
85. Purschke M, Rubio N, Held KD, Redmond RW (2010) Phototoxicity of hoechst 33342 in time-lapse fluorescence microscopy. Photochem Photobiol Sci 9:1634–1639
86. Jones SA, Shim S-H, He J, Zhuang X (2011) Fast, three-dimensional super-resolution imaging of live cells. Nat Methods 8:499–508
87. Huang B, Bates M, Zhuang X (2009) Super-resolution fluorescence microscopy. Annu Rev Biochem 78:993–1016

10 Mathematical Chemodescriptors and Biodescriptors: Background and Their Applications in the Prediction of Bioactivity/Toxicity of Chemicals

Subhash C. Basak

10.1 Introduction

At quite uncertain times and places,
The atoms left their heavenly path,
And by fortuitous embraces,
Engendered all that being hath.
And though they seem to cling together,
And form 'associations' here,
Yet, soon or late, they burst their tether,
And through the depths of space career.

– James Clerk Maxwell
In: "Molecular Evolution," Nature, 8, 1873.
In Lewis Campbell and William Garnett, The Life of James Clerk Maxwell (1882), 637

Many physiological, pathological, toxicological, and biomedicinal processes are determined by interactions of small molecules such as endogenous ligands, drugs, xenobiotics, and substrates as well as inhibitors of enzymes related to metabolic pathways with their appropriate biological targets. The maintenance of the integrity and continuity of such key ligand-biotarget interactions is critical for the smooth functioning of biological systems ranging from the single-celled organism to the complex ecosystems. A large number of drugs are small molecules that interact with specialized enzymes/receptors in appropriate physiological compartments and thereby produce effect(s) that bring a pathologically perturbed biological system back to a healthy state [1–4]. Biological properties of molecules, beneficial or deleterious, can be looked upon as the result of ligand-biotarget interactions and can be expressed by the relationship:

$$BR = f\left(S, B\right) \tag{10.1}$$

where *BR* represents the normal biological or pathological/toxicological response produced by the ligand (drug or toxicant) in the target biological system and *B* represents the relevant biochemical part of the target system which is perturbed by ligand to produce the measurable effect. It is believed that a major determinant of *BR* is the nature or structure (S) of the ligand. The structure becomes the sole determinant of the variation of the measured BR from one chemical to another when the biological system, *B*, remains practically the same during the course of the experiment and there is alternation only in the structure of the ligands. Eq. 10.1 under such a condition approximates to:

$$BR = f\left(S\right) \tag{10.2}$$

A lot of research conducted in drug discovery, toxicology, environmental sciences, and biochemistry follows the paradigm expressed in Eq. 10.2, and using this relationship researchers

S.C. Basak (✉)
International Society of Mathematical Chemistry, University of Minnesota Duluth-Natural Resources Research Institute, Duluth, MN, USA

Department of Chemistry and Biochemistry, University of Minnesota Duluth, 5013 Miller Trunk Highway, Duluth, MN 55811, USA
e-mail: sbasak@nrri.umnj.edu

S. Singh (ed.), *Systems Biology Application in Synthetic Biology*,
DOI 10.1007/978-81-322-2809-7_10

attempt to decipher the effects as well as the modes and mechanism(s) of action of molecules on some selected biotargets, which are assumed not to change significantly during the course of the experiment.

When we embark on the characterization of *BR* based on chemical structure alone following Eq. 10.2, we really attempt to understand which characteristics of the chemical structure are recognized by the biomolecular target. What are the factors involved in recognition: molecular size, shape, chirality, stereo-electronic nature, or charge? Which ones are more important and which have a marginal impact on *BR*? This is often accomplished by the development of molecular descriptors, referred to by us as chemodescriptors, which quantify various aspects of molecular structure such as shape, size, symmetry, chirality, stereo-electronic nature, etc. using various mathematical techniques.

10.2 Mathematical Characterization of Structure: Molecules and Biomolecules

> Ostensibly there is color, ostensibly sweetness, ostensibly bitterness, but actually only atoms and the void.
>
> Galen
> In: Nature and the Greeks, Erwin Schrodinger, 1954

> In order to describe an aspect of holistic reality we have to ignore certain factors such that the remainder separates into facts. Inevitably, such a description is true only within the adopted partition of the world, that is, within the chosen context.
>
> Hans Primas
> Chemistry, Quantum Mechanics and Reductionism [5]

10.2.1 The Molecular Structure Conundrum: Simple Graph to Quantum Chemical Hamiltonians

The structure of an assembled entity is the pattern of relationship among its parts. Molecular structure can be looked upon as the representation of the relationship among its various constituents. The term *molecular structure* represents a set of nonequivalent and probably disjoint concepts [5]. There is no reason to believe that when we discuss diverse topics, e.g., chemical synthesis, reaction rates, spectroscopic transitions, chemical reaction mechanisms, and *ab initio* calculations, using the notion of molecular structure, the different meanings we attach to the single term "molecular structure" originate from the same fundamental concept [6, 7]. In the context of molecular science, the various concepts of molecular structure, e.g., classical valence bond representations, various chemical graph theoretic representations, ball and spoke model of a molecule, representation of a molecule by minimum energy conformation, and representation of chemical species by Hamiltonian operators, are model objects [8–15] derived through different abstractions of the same chemical reality. In each instance, the *equivalence class* (concept or model of molecular structure) is generated by selecting certain aspects while ignoring some unique properties of those actual entities. This explains the plurality of the concept of molecular structure and their autonomous nature, the word "autonomous" being used here in the same sense that one concept is not logically derived from the other [7].

10.2.2 The Philosophical Basis of Modeling in Mathematical Chemistry

The process of modeling arises out of abstraction from sense data derived from reality. As put forward by Albeit Einstein [8] in his remarks on the philosopher Bertrand Russell's theory of knowledge:

> The more, however, we turn to the most primitive concepts of everyday life, the more difficult it becomes amidst the mass of inveterate habits to recognize the concept as an independent creation of thinking. It was thus that the fateful conception -fateful, that is to say, for an understanding of the here-existing conditions – could arise, according to which the concepts originate from experience by way of "abstraction," i.e., through omission of a part of its content.

As pointed out by Basak [8] regarding the philosophy of modeling [9] of molecular structure:

> Any concept of molecular structure is a hypothetical sketch of the organization of molecules. Such a model object is a general theory and remains empirically untestable. A model object has to be grafted onto a specific theory to generate a theoretical model. A theoretical model of an object can be empirically tested. For example, when it was suggested by Sylvester [12] in 1878 that the structural formula of a molecule is a special kind of graph, it was an innovative general theory without any predictive potential. When the idea of combinatorics was applied on chemical graphs (model objects), it could be predicted that "there should be exactly two isomers of butane (C_4H_{10})" because "there are exactly two tree graphs with four verüces" when one considers only the non-hydrogen atoms present in C_4H_{10}. This is a theoretical model of limited predictive potential. Although it predicts the existence of chemical species, given a set of molecules, e.g. isomers of hexane (C_6H_{14}), the model is incapable of predicting any property. This is because of the fact that any empirical property P maps a set of chemical structures into the set R of real numbers and thereby orders the set empirically. Therefore, to predict the property from structure, we need a nonempirical (structural) ordering scheme which closely resembles the empirical ordering of structures as determined by P. This is a more specific theoretical model based on the same model object (chemical graph) and can be accomplished by using specific graph invariant(s).

10.2.3 Mathematical Chemodescriptors: Topological Indices, 3D Descriptors, and Quantum Chemical Indices

One of the important goals of structural chemistry, biomedicinal chemistry, and computational toxicology is the "optimal characterization" of molecular structure for the purpose of predicting their properties. As discussed in Sect. 10.2.1, optimal characterization of structure has remained elusive. Different groups of researchers have used different methods for the representation and quantification of molecular structure. In our quantitative structure-activity relationship (QSAR) and quantitative molecular similarity analysis (QMSA) research, we have used mainly three classes of descriptors for the quantification of structure, viz., (a) graph invariants defined on molecular graphs, also known as *topological indices*, (b) three-dimensional (3D) or geometrical descriptors, and (c) quantum chemical descriptors.

In our research, we have also used *atom pairs* (APs), which are fragment-based descriptors. The method of Carhart et al. [10] was used to calculate the atom pairs, which defines an atom pair as a substructure consisting of two non-hydrogen atoms *i* and *j* and their interatomic separation:

$$\langle \text{atom descriptor}_i \rangle - \langle \text{separation} \rangle - \langle \text{atom descriptor}_j \rangle$$

where <atom descriptor> contains information regarding atom type, number of non-hydrogen neighbors and the number of π electrons. The interatomic separation is defined as the number of atoms traversed in the shortest bond-by-bond path containing both atoms.

Graph theory was discovered by Euler [11] in 1736. Sylvester [12] in 1878 saw the clear-cut relationship between graph theory and molecular structure. He also commented on the connection between chemistry and mathematics in general, as evident from the following [13]:

> Chemistry has the same quickening and suggestive influence upon the algebraist as a visit to the Royal Academy, or the old masters may be supposed to have on a Browning or a Tennyson. Indeed it seems to me that an exact homology exists between painting and poetry on the one hand and modem chemistry and modem algebra on the other. In poetry and algebra we have the pure idea elaborated and expressed through the vehicle of language, in painting and chemistry the idea is enveloped in matter, depending in part on manual processes and the resources of art for its due manifestation.

Applications of graph theory to chemical problems are part of a fast developing field of science called mathematical chemistry or, more correctly, discrete mathematical chemistry. Although Sylvester [12] saw the connection between molecular structure and chemistry as back as 1878, modern research in chemical graph theory had its humble beginning at the middle of the twentieth century probably with the publication of the seminal paper by Harry Wiener [14] on the calculation of structural indices for the prediction of molecular properties. Invariants of graphs associated with molecules and biomolecules quantify certain aspects of their structure and have been used in the characterization and comparison of such structures as well as prediction of their properties Specifically, such invariants and orthogonal factors like *principal components* (PCs) derived from them have found applications in QSAR studies [15–18], QMSA research [18–22], clustering of large libraries of structures into smaller subsets [20, 21], and in the discrimination of pathological structures like isospectral graphs [15].

The author of this chapter (Basak) and his coworkers have been involved since the early 1970s in the development of novel numerical graph invariants or topological indices (TIs) [16–19, 22–26] as well as biodescriptors derived from DNA/RNA sequences [16, 27] and proteomics maps [28]. It may be mentioned here that graph theoretical numerical indices were called "topological indices" by Hosoya [29] for the first time in a paper published in 1971.

Many topological indices can be conveniently derived from various matrices including the *adjacency matrix A* (*G*) and the *distance matrix D* (*G*) of a *chemical graph G*. These matrices are usually constructed from labeled graphs of hydrogen-suppressed molecular skeletons. For details of theoretical basis and calculation of topological indices, see refs [17, 18, 23–29].

Basak et al. have divided the topological indices (TIs) into two major groups: topostructural (TS) indices and topochemical (TC) indices. TS indices are calculated from skeletal graph models of molecules which do not distinguish among different types of atoms in a molecule or the various types of chemical bonds, e.g., single bond, double bond, triplet bond, etc. Thus, TS indices quantify information regarding the connectivity, adjacency, and distances between vertices ignoring their distinct chemical nature. TC indices, on the other hand, are sensitive to both the pattern of connectedness of the vertices (atoms), as well as their chemical bonding characteristics. Therefore, the TC indices are more complex and chemically informative as compared to the TS descriptors.

The geometrical or 3D parameters quantify the volume, size, and shape of molecules from various models. We have used van der Waals' volume as a measure of gross size of molecules. The three-dimensional Wiener indices calculated on the hydrogen-suppressed and hydrogen-filled graphs are also quantifiers of molecular shape and size. With respect to calculation of quantum chemical descriptors, we have used both the AM_1 semiempirical method and *ab initio* calculations based on the STO-3G, 6-31G(d), 6-311G, 6-311G(d), and aug-cc-pVTZ basis sets. For chemodescriptors used by Basak group in their studies, see [18, 29–35]. Table 10.1 gives the symbols and definition of molecular chemodescriptors.

10.2.4 Hierarchical Classification of Descriptors

The combination of topological, geometrical, and quantum chemical chemodescriptors, and biodescriptors (*vide infra*) derived from proteomics, genomics, and DNA sequence characterization, leads to a hierarchy of descriptors that begins with the simplest graph invariants and ends with the biodescriptors, which require expensive and time-intensive laboratory test data (Fig. 10.1). It should be clearly stated here that descriptors in the higher levels of the hierarchy are not necessarily superior to those placed at lower levels. The scheme simply shows a gradation based on the need for computational and laboratory resources.

The molecular descriptors itemized in Table 10.1 are calculated by Basak's team using Molconn-Z [30], POLLY [31], APProbe [32], and Triplet [33], MOPAC [34], and Gaussian [35].

10.3 Quantitative Structure-Activity Relationship (QSAR) Using Chemodescriptors

Those alone are wise who act after investigation.

Charaka
In Sutrasthana, 10:5

We haven't got the money, so we've got to think

Ernest Rutherford

Table 10.1 Symbols, definitions, and classification of structural molecular descriptors

	Topostructural (TS)
I_D^W	Information index for the magnitudes of distances between all possible pairs of vertices of a graph
$\overline{I_D^W}$	Mean information index for the magnitude of distance
W	Wiener index = half-sum of the off-diagonal elements of the distance matrix of a graph
I^D	Degree complexity
H^V	Graph vertex complexity
H^D	Graph distance complexity
$\overline{IC}$	Information content of the distance matrix partitioned by frequency of occurrences of distance *h*
M_1	A Zagreb group parameter = sum of square of degree over all vertices
M_2	A Zagreb group parameter = sum of cross-product of degrees over all neighboring (connected) vertices
$^h\chi$	Path connectivity index of order $h=0–10$
$^h\chi_C$	Cluster connectivity index of order $h=3–6$
$^h\chi_{PC}$	Path-cluster connectivity index of order $h=4–6$
$^h\chi_{Ch}$	Chain connectivity index of order $h=3–10$
P_h	Number of paths of length $h=0–10$
J	Balaban's *J* index based on topological distance
nrings	Number of rings in a graph
ncirc	Number of circuits in a graph
DN^2S_y	Triplet index from distance matrix, square of graph order, and distance sum; operation $y=1–5$
DN^21_y	Triplet index from distance matrix, square of graph order, and number 1; operation $y=1–5$
$AS1_y$	Triplet index from adjacency matrix, distance sum, and number 1; operation $y=1–5$
$DS1_y$	Triplet index from distance matrix, distance sum, and number 1; operation $y=1–5$
ASN_y	Triplet index from adjacency matrix, distance sum, and graph order; operation $y=1–5$
DSN_y	Triplet index from distance matrix, distance sum, and graph order; operation $y=1–5$

(continued)

Table 10.1 (continued)

	Topostructural (TS)
DN^2N_y	Triplet index from distance matrix, square of graph order, and graph order; operation $y=1–5$
ANS_y	Triplet index from adjacency matrix, graph order, and distance sum; operation $y=1–5$
$AN1_y$	Triplet index from adjacency matrix, graph order, and number 1; operation $y=1–5$
ANN_y	Triplet index from adjacency matrix, graph order, and graph order again; operation $y=1–5$
ASV_y	Triplet index from adjacency matrix, distance sum, and vertex degree; operation $y=1–5$
DSV_y	Triplet index from distance matrix, distance sum, and vertex degree; operation $y=1–5$
ANV_y	Triplet index from adjacency matrix, graph order, and vertex degree; operation $y=1–5$
Topochemical (TC)	
O	Order of neighborhood when IC_r reaches its maximum value for the hydrogen-filled graph
O_{orb}	Order of neighborhood when IC_r reaches its maximum value for the hydrogen-suppressed graph
I_{ORB}	Information content or complexity of the hydrogen-suppressed graph at its maximum neighborhood of vertices
IC_r	Mean information content or complexity of a graph based on the *r*th ($r=0–6$) order neighborhood of vertices in a hydrogen-filled graph
SIC_r	Structural information content for *r*th ($r=0–6$) order neighborhood of vertices in a hydrogen-filled graph
CIC_r	Complementary information content for *r*th ($r=0–6$) order neighborhood of vertices in a hydrogen-filled graph
$^h\chi^b$	Bond path connectivity index of order $h=0–6$
$^h\chi^b_C$	Bond cluster connectivity index of order $h=3–6$
$^h\chi^b_{Ch}$	Bond chain connectivity index of order $h=3–6$
$^h\chi^b_{PC}$	Bond path-cluster connectivity index of order $h=4–6$
$^h\chi^v$	Valence path connectivity index of order $h=0–6$
$^h\chi^v_C$	Valence cluster connectivity index of order $h=3–6$

(continued)

Table 10.1 (continued)

	Topostructural (TS)
${}^{h}\chi^{v}{}_{Ch}$	Valence chain connectivity index of order $h=3–6$
${}^{h}\chi^{v}{}_{PC}$	Valence path-cluster connectivity index of order $h=4–6$
J^B	Balaban's *J* index based on bond types
J^X	Balaban's *J* index based on relative electronegativities
J^Y	Balaban's *J* index based on relative covalent radii
AZV_y	Triplet index from adjacency matrix, atomic number, and vertex degree; operation $y=1–5$
AZS_y	Triplet index from adjacency matrix, atomic number, and distance sum; operation $y=1–5$
ASZ_y	Triplet index from adjacency matrix, distance sum, and atomic number; operation $y=1–5$
AZN_y	Triplet index from adjacency matrix, atomic number, and graph order; operation $y=1–5$
ANZ_y	Triplet index from adjacency matrix, graph order, and atomic number; operation $y=1–5$
DSZ_y	Triplet index from distance matrix, distance sum, and atomic number; operation $y=1–5$
DN^2Z_y	Triplet index from distance matrix, square of graph order, and atomic number; operation $y=1–5$
nvx	Number of non-hydrogen atoms in a molecule
nelem	Number of elements in a molecule
fw	Molecular weight
${}^{h}\chi^{v}$	Valence path connectivity index of order $h=7–10$
${}^{h}\chi^{v}{}_{Ch}$	Valence chain connectivity index of order $h=7–10$
si	Shannon information index
totop	Total topological index *t*
sumI	Sum of the intrinsic state values *I*
sumdelI	Sum of delta-*I* values
tets2	Total topological state index based on electrotopological state indices
phia	Flexibility index (kp_1* kp_2/nvx)
Idcbar	Bonchev-Trinajstić information index
IdC	Bonchev-Trinajstić information index
Wp	Wienerp
Pf	Plattf
Wt	Total Wiener number
knotp	Difference of chi-cluster-3 and path-cluster-4
knotpv	Valence difference of chi-cluster-3 and path-cluster-4

(continued)

Table 10.1 (continued)

	Topostructural (TS)
nclass	Number of classes of topologically (symmetry) equivalent graph vertices
NumHBd	Number of hydrogen bond donors
NumHBa	Number of hydrogen bond acceptors
SHCsats	E-State of C sp^3 bonded to other saturated C atoms
SHCsatu	E-State of C sp^3 bonded to unsaturated C atoms
SHvin	E-State of C atoms in the vinyl group, =CH-
SHtvin	E-State of C atoms in the terminal vinyl group, $=CH_2$
SHavin	E-State of C atoms in the vinyl group, =CH-, bonded to an aromatic C
SHarom	E-State of C sp^2 which are part of an aromatic system
SHHBd	Hydrogen bond donor index, sum of hydrogen E-State values for –OH, =NH, $-NH_2$, -NH-, -SH, and #CH
SHwHBd	Weak hydrogen bond donor index, sum of CH hydrogen E-State values for hydrogen atoms on a C to which a F and/or Cl are also bonded
SHHBa	Hydrogen bond acceptor index, sum of the E-State values for –OH, =NH, $-NH_2$, -NH-, >N-, -O-, -S-, along with –F and –Cl
Qv	General polarity descriptor
$NHBint_y$	Count of potential internal hydrogen bonders ($y=2–10$)
$SHBint_y$	E-State descriptors of potential internal hydrogen bond strength ($y=2–10$)
	Electrotopological state index values for atoms types:
	SHsOH, SHdNH, SHsSH, SHsNH2, SHssNH, SHtCH, SHother, SHCHnX, Hmax Gmax, Hmin, Gmin, Hmaxpos, Hminneg, SsLi, SssBe, Sssss, Bem, SssBH, SsssB, SssssBm, SsCH3, SdCH2, SssCH2, StCH, SdsCH, SaaCH, SsssCH, SddC, StsC, SdssC, SaasC, SaaaC, SssssC, SsNH3p, SsNH2, SssNH2p, SdNH, SssNH, SaaNH, StN, SsssNHp, SdsN, SaaN, SsssN, SddsN, SaasN, SssssNp, SsOH, SdO, SssO, SaaO, SsF, SsSiH3, SssSiH2, SsssSiH, SssssSi, SsPH2, SssPH, SsssP, SdsssP, SsssssP, SsSH, SdS, SssS, SaaS, SdssS, SddssS, SssssssS, SsCl, SsGeH3, SssGeH2, SsssGeH, SssssGe, SsAsH2, SssAsH, SsssAs, SdsssAs, SsssssAs, SsSeH, SdSe, SssSe, SaaSe, SdssSe, SddssSe, SsBr, SsSnH3, SssSnH2, SsssSnH, SssssSn, SsI, SsPbH3, SssPbH2, SsssPbH, SssssPb
Geometrical (3D)/shape	
kp_0	Kappa zero

(continued)

Table 10.1 (continued)

	Topostructural (TS)
kp_1-kp_3	Kappa simple indices
ka_1-ka_3	Kappa alpha indices
V_W	Van der Waals volume
^{3D}W	3D Wiener number based on the hydrogen-suppressed geometric distance matrix
$^{3D}W_H$	3D Wiener number based on the hydrogen-filled geometric distance matrix
	Quantum chemical (QC)
E_{HOMO}	Energy of the highest occupied molecular orbital
E_{HOMO-1}	Energy of the second highest occupied molecular
E_{LUMO}	Energy of the lowest unoccupied molecular orbital
E_{LUMO+1}	Energy of the second lowest unoccupied molecular orbital
ΔHf	Heat of formation
μ	Dipole moment

Modern society routinely uses a large number of natural and man-made chemicals in the form of drugs, solvents, synthetic intermediates, cosmetics, herbicides, pesticides, etc. to maintain the lifestyle. But in many cases, a large fraction of these chemicals do not have the experimental data necessary for the prediction of their beneficial and deleterious effects [36]. Table 10.2 gives a partial list of properties, both physical and biochemical/pharmacological/toxicological, needed for the effective screening of chemicals for new drug discovery and protection of human as well as ecological health. Because determination of such properties for so many chemicals in the laboratory is prohibitively costly, one solution of this quagmire has been the use of QSARs and molecular similarity-based analogs to obtain acceptable estimated values of properties.

10.3.1 Statistical Methods for QSAR Model Development and Validation

In God we trust. All others must bring data.

W. Edwards Deming

To call in the statistician after the experiment is done maybe no more
than asking him to perform a post-mortem examination:
he may be able to say what the experiment died of.

Ronald Fisher:
http://www.brainyquote.com/quotes/authors/r/ronald_fisher.html

In the early 1970s, when this author (Basak) started carrying out research on the development and use of calculated chemodescriptors in QSAR, only a few such descriptors were available. But now, with the availability of various software [30–35, 37, 38], the landscape of availability and calculation of molecular descriptors is very different. The four major pillars [18] of a useful QSAR system development are:

(a) Availability of high-quality experimental data (veracity of dependent variable)
(b) Data on sufficient number of compounds (volume or reasonably good sample size)
(c) Availability of relevant descriptors (independent variables of QSAR) which quantify aspects of molecular structure relevant to the activity/toxicity of interest
(d) Use of appropriate methods for model building and validation

The various pathways for the development of structure-activity relationship (SAR) and property-activity relationship (PAR) models either from calculated molecular descriptors or from experimentally determined as well as calculated properties as independent variables may be expressed by the scheme provided in Fig. 10.2.

The use of computed molecular descriptors and experimental property data in PAR/SAR/QSAR may be illuminated through a formal exposition of the structure-property similarity principle – the central paradigm of the field of SAR [39]. Figure 10.2 depicts the determination of an experimental property, e.g., measurement of octanol-water partition coefficient of a chemical in the laboratory, as a function $\alpha: C \rightarrow R$ which maps the set C of compounds into the real line R. A nonempirical QSAR may be looked upon as a composition of a description function $\beta_1: C \rightarrow D$ mapping each chemical structure of C

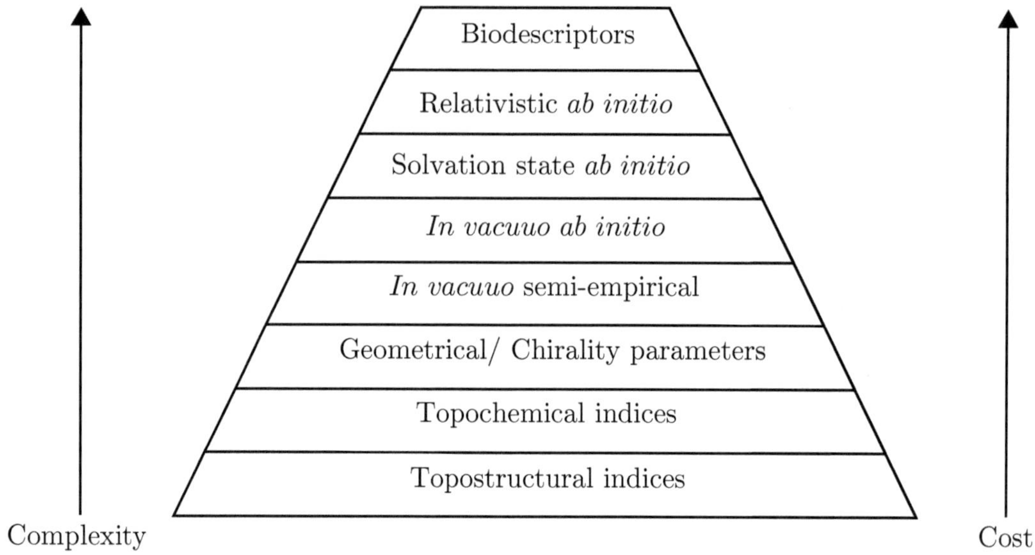

Fig. 10.1 Hierarchical classification of chemodescriptors and biodescriptors used in QSAR (Source: Basak [18]. With permission from Bentham Science Publishers)

Table 10.2 List of properties needed for screening of chemicals

Physicochemical	Pharmacological/toxicological
Molar volume	Macromolecular level
Boiling point	Receptor binding (K_d)
Melting point	Michaelis constant (K_m)
Vapor pressure	Inhibitor constant (K_i)
Water solubility	DNA alkylation
Dissociation constant (pKa)	Unscheduled DNA synthesis
Partition coefficient	Cell level
Octanol-water (log P)	Salmonella mutagenicity
Air-water	Mammalian cell transformation
Sediment-water	Organism level (acute)
Reactivity (electrophilicity)	LD_{50} (mouse, rat)
	LC_{50} (fathead minnow)
	Organism level (chronic)
	Bioconcentration factor
	Carcinogenicity
	Reproductive toxicity
	Delayed neurotoxicity
	Biodegradation

into a space of nonempirical structural descriptors (D) and a prediction function β_2: D → R which maps the descriptors into the real line. One example can be the use of Molconn-Z [30] indices for the development of QSARs. When [α(C) – $\beta_2 \circ \beta_1$ (C)] is within the range of experimental errors, we say that we have a good QSAR model. On the other hand, PAR is the composition of θ_1: C → M which maps the set C into the molecular property space M and θ_2: M → R mapping those molecular properties into the real line R. Property-activity relationship seeks to predict one property (usually a complex physicochemical property) or bioactivity of a molecule in terms of other (usu-

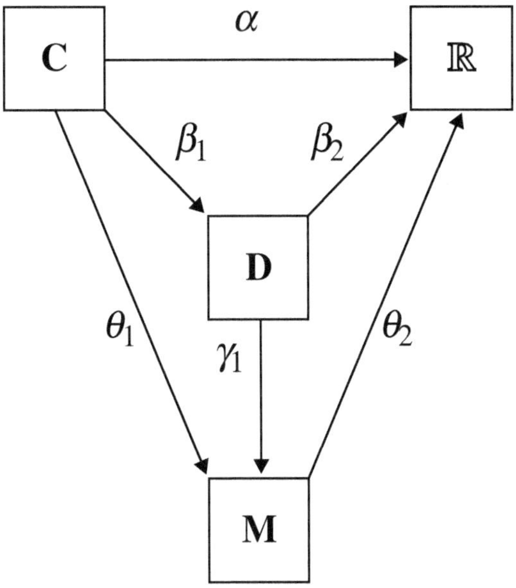

Fig. 10.2 Composition functions of various mappings for structure-activity relationship (SAR) and property-activity relationship (PAR) (Source: Basak and Majumdar [46]. With permission from Bentham Science Publishers)

ally simpler or easily determined experimentally) properties.

Basak group uses the following generic method in the validation of QSAR models: In the process of formulating a scientifically interpretable and technically sound QSAR model, we need to keep in mind some important issues. First and foremost, one has to check whether a specific method is the best technique in modeling a specific QSAR scenario. In a regression set up, for example, when the number of independent variables or descriptors (p) is much larger than the number of data points (dependent variable, n), i.e., $p>>n$, the estimate of the coefficient vector is nonunique. This is also the case when predictors in the study are highly correlated with one another to the extent that the "design matrix" is rank-deficient. Both of these factors are relevant to QSARs. In many contemporary QSAR studies, the number of initial predictors typically is in the range of hundreds or thousands, whereas more often than not, mostly to keep cost of generation of experimental data under control, the experimenter can collect data on only a much smaller number (tens or hundreds) of samples. This effectively makes the problem high dimensional and rank-deficient ($p>>n$) in nature. Also, when a large number of descriptors on a set of chemicals are used to model their activity, one should expect that some predictors within a single class, e.g., TC descriptors, or even predictors belonging to apparently different classes are highly correlated with one another. Such situations can be tackled either by attempting to pick important variables through model selection or "sparsity"-type approaches (e.g., forward selection, LASSO [40], adaptive LASSO [41]), or finding a lower-dimensional transformation that preserves most of the information present in the set of descriptors, e.g., principal component analysis (PCA) and envelope methods [42].

We need to check the ability of a model to give competent predictions on "similar" data sets via validation on out-of-sample test sets. For a relatively small sample, i.e., a small set of compounds, this is achieved by carrying out a **leave-one-out (LOO) cross-validation**. For data sets with a large number of compounds, a more computationally economical way is to do a **k-fold cross-validation**: split the data set randomly into k (previously decided by the researcher) equal subsets, take each subset in turn as test set, and use the remaining compounds as training sets and use the model to obtain predictions. Comparing cross-validation with the somewhat prevalent approach in QSAR research of **external validation**, i.e., choosing a single train-test split of compounds, it should be pointed out that in external validation, the splits of data sets are carried out only once using the experimenters' *a priori* knowledge or some subjectively chosen ad hoc criterion. But in cross-validation, the splits are chosen randomly, thus providing a more unbiased estimate of the generalizability of the QSAR model. Furthermore, Hawkins et al. [43] proved theoretically that compared to external validation, LOO cross-validation is a better estimator of the actual predictive ability of a statistical model for small data sets, while for large sample size both perform equally well. To quote Hawkins et al. [43], "The bottom line is that in the typical QSAR setting where available sample sizes are modest, holding back compounds for model testing is ill-advised. This fragmentation of the sample harms the calibration and does not give a trustworthy assessment of fit anyway. It is better to use all data for the calibration step and check the fit by cross-validation, making sure that the cross-validation is carried out correctly." Specific drawbacks of holding out only one test set in the external validation method include: (1) structural features of the held out chemicals are not included in the modeling process, resulting in a loss of information; (2) predictions are made on only a subset of the available compounds, whereas the LOO method predicts the activity value for all compounds; (3) there is no scientific tool that can guarantee similarity between chemicals in the training and test sets; and (4) personal bias can easily be introduced in selection of the external test set.

In the rank-deficient situation of QSAR formulation, special care should be taken in combining conventional modeling with the additional step of variable selection or dimension reduction. An intuitive, but frequently misunderstood and wrong, procedure would be to perform the first stage of preprocessing first, selecting important variables or determining the optimal transformation, and then use the transformed data/selected variables to build the predictive QSAR models and obtain predictions for each train-test split. The reason why this is not appropriate is that the data is split only after the variable selection/dimension reduction step is already completed. Essentially this method ends up using information from the holdout compound/split subset to predict activity of those very samples. This *naïve cross-validation* procedure causes synthetic inflation of the cross-validated q^2, hence compromises the predictive ability of the model [44, 45] (Fig. 10.3). A two-step procedure (referred in Fig. 10.3 as *two-deep CV*) helps avoid this tricky situation. Instead of doing the pre-model building step first and then taking multiple splits for out-of-sample prediction, for each split of the data the initial steps are performed only using the training set of compounds each time. Since calculations on two different splits are not dependent on each other, for large data sets the increased computational demand arising out of the repeated variable selection can be tackled using substantial computer resources like parallel processing. It should be emphasized that the naïve cross-validation (naïve CV) method gives **naïve or wrong q^2** values, whereas the two-deep cross-validation (two-deep CV) approach gives us the correct or **true q^2**.

For recent reviews and research on this topic of proper cross-validation, please see the recent publications of Basak and coworkers [46–52].

The quality of the model, in terms of its predictive ability, is evaluated based on the associated q^2 value, which is defined as:

$$q^2 = 1 - (\text{PRESS}/\text{SSTotal}) \quad (10.3)$$

where PRESS is the prediction sum of squares and SSTotal is the total sum of squares. Unlike R^2 which tends to increase upon the addition of any descriptor, q^2 will decrease upon the addition of irrelevant descriptors, thereby providing a reliable measure of model quality.

In order to illustrate practically the inflation of q^2 associated with the use of improper statistical techniques, we deliberately developed a wrong model using stepwise ordinary least squares (OLS) regression, which is commonly used in many QSAR studies but often results in overfitting and renders the model unreliable for making predictions for chemicals similar to those used to calibrate the model. The REG procedure of the SAS statistical package [53] was used to develop stepwise regression model. For details see [45]. Rat fat/air partition coefficient values for a diverse set of 99 organic compounds were used for this study. It should be noted that two compounds with fewer than three non-hydrogen atoms, for which we could not calculate our entire suite of structure-based descriptors, were omitted from our study. A total of 375 descriptors were calculated using software packages including POLLY v2.3, Triplet, Molconn-Z v 3.5, and Gaussian 03W v6.0. This is clearly a rank-deficient case with the number of compounds ($n=97$) being much smaller than the number of predictors ($p=375$). The ridge regression (RR) approach [45, 51] in which the Gram-Schmidt algorithm was used to properly thin the descriptors yielded a four-parameter model with an associated q^2 of 0.854. Each of the four descriptors was topological in nature; none of the three-dimensional or quantum chemical descriptors were selected. An inflated q^2 of 0.955 was

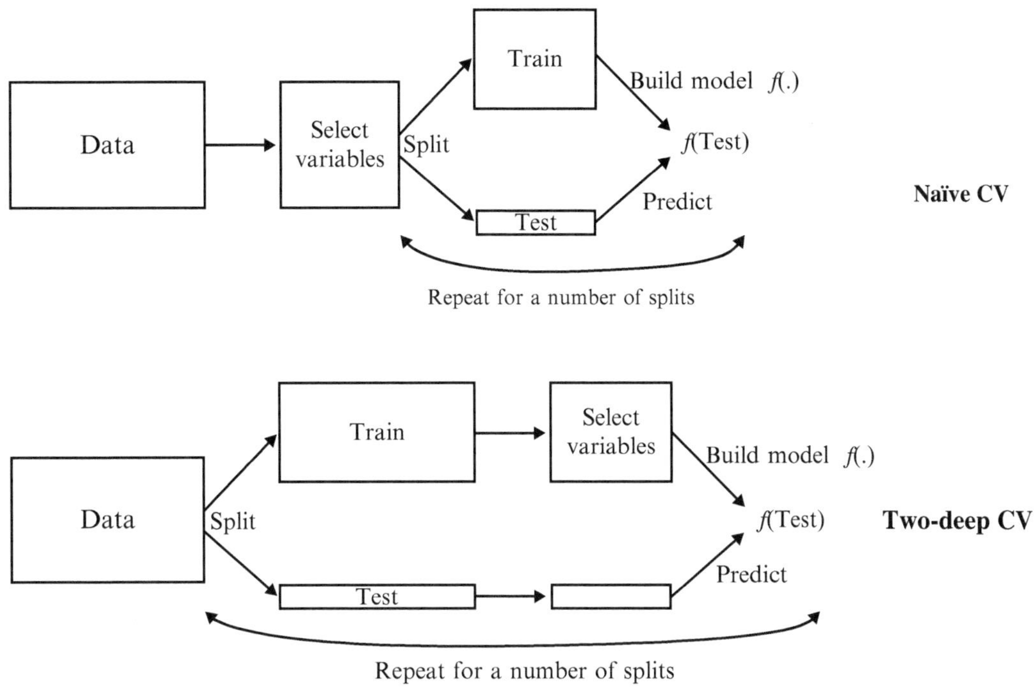

Fig. 10.3 Difference between naïve and two-deep cross-validation (CV) schemes (Source: Basak and Majumdar [46]. With permission from Bentham Science Publishers)

obtained from the stepwise regression approach which yielded a 24-parameter model.

10.3.2 Intrinsic Dimensionality of Descriptor Spaces: Use of Principal Component Analysis (PCA) as the Parsimony Principle or Occam's Razor

शैले शैले न माणिक्यं मौक्तिकं न गजे गजे ।
साधवो न हि सर्वत्र चंदनं न वने वने ॥

shaile shaile na maanikyam mauktikam na gaje gaje
saadhavo naahi sarvatra chandanam na vane vane
(In Sanskrit)

Not all mountains contain gems in them, nor does every elephant has pearl in it, noble people are not found everywhere, nor is sandalwood found in every forest.

Chanakya

You gave too much rein to your imagination. Imagination is a good servant, and a bad master. The simplest explanation is always the most likely.
– Agatha Christie

As discussed earlier, these days we can calculate a large number of molecular descriptors using the available software. But **all descriptors are not created equal and each descriptor is not needed for all modeling situations**. In the QSAR scenario, we need to use proper methods for the selection of relevant descriptors. Methods like principal component analysis (PCA) [19, 54, 55] and interrelated two-way clustering (ITC) [56] can be used for variable selection or descriptor thinning.

When p molecular descriptors are calculated for n molecules, the data set can be viewed as n vectors in p dimensions, each chemical being represented as a point in R^p. Because many of the descriptors are strongly correlated, the n points in R^p will lie on a subspace of dimension lower than p. Methods like principal component analysis can be used to characterize the *intrinsic dimensionality* of chemical spaces. Since the early 1980s, Basak and coworkers have carried out PCA of various congeneric and diverse data sets relevant to new drug discovery and predictive toxicology. Principal components (PCs) derived from mathematical chemodescriptors have been used in the formulation of quantitative structure-activity relationships (QSARs), clustering of large combinatorial libraries, as well as quantitative molecular similarity analysis (QMSA), the last one to be discussed later. This section of the article will discuss PCA studies on characterization and visualization of chemical spaces of two data sets, one congeneric and one structurally diverse: (1) a large and structurally diverse set of 3692 chemicals which was a subset of the Toxic Substances Control Act (TSCA) Inventory maintained by the US Environmental Protection Agency (USEPA) and (2) a virtual library of 248,832 psoralen derivatives,

In the early 1980s, after Basak joined the University of Minnesota Duluth, the software POLLY [31] was developed and large-scale calculation of TIs for QSAR and QMSA analyses was initiated. In one of the earliest studies of its kind, Basak et al. [19, 57] used the first version of POLLY for the calculation of 90 TIs for a collection of 3692 structurally diverse chemicals which was a subset of the Toxic Substances Control Act (TSCA) Inventory of USEPA. The authors carried out PCA on this data set and asked the question: **What is the intrinsic dimensionality of chemical structure measured by the large number of TIs**? As shown in the summary in Table 10.3, first ten PCs with eigenvalues greater than or equal to 1.0 explained 92.6 % of the variance in the data of the calculated descriptors, and first four PCs explained 78.3 % of the variance [19, 57]. For a recent review of our research in this line, see Basak et al. [58].

It is clear from the data in Table 10.3 that PC_1 is strongly correlated with those indices which are related to the size of chemicals. It is noteworthy that for the set of 3692 diverse chemicals PC_1 was also highly correlated with molecular weight ($r=0.81$) and K_0 (0.95) which is the number of vertices in hydrogen-suppressed graphs. PC_2 was interpreted by us as an axis of molecular complexity as encoded by the higher-order information theoretic indices developed by Basak group [23, 59]. PC_3 is most highly related to the cluster/path-cluster-type molecular connectivity indices which quantify structural aspects regarding molecular branching. The data in Table 10.3 clearly show that PC_4 is strongly correlated with the cyclicity terms of the connectivity class of topological indices [19].

Table 10.3 Correlation of the first four PCs with the original variables in the 90 topological indices, [19, 57]

PC_1	PC_2	PC_3	PC_4
K_1 (0.96)	SIC_3 (0.97)	${}^4\chi^b{}_C$ (0.69)	${}^4\chi_{CH}$ (0.85)
${}^2\chi$ (0.95)	CIC_4 (−0.96)	${}^4\chi^b{}_C$ (0.69)	${}^4\chi^b{}_{CH}$ (0.84)
${}^3\chi$ (0.95)	CIC_3 (−0.95)	${}^5\chi^b{}_C$ (0.68)	${}^4\chi^v{}_{CH}$ (0.80)
K_2 (0.95)	SIC_4 (0.95)	${}^4\chi_C$ (0.68)	${}^3\chi_{CH}$ (0.75)
K_0 (0.95)	SIC_2 (0.94)	$3\chi v_C$ (0.67)	${}^3\chi^b{}_{CH}$ (0.75)
${}^1\chi$ (0.94)	CIC_5 (−0.94)	${}^5\chi_C$ (0.64)	${}^4\chi^b{}_{CH}$ (0.74)
${}^3\chi^b$ (0.94)	CIC_6 (−0.92)	${}^6\chi_C$ (0.64)	${}^3\chi^v{}_{CH}$ (0.72)
${}^4\chi$ (0.94)	SIC_5 (0.92)	${}^3\chi_C$ (0.61)	${}^5\chi_{CH}$ (0.71)
${}^4\chi^b$ (0.93)	SIC_6 (0.89)	${}^6\chi^b{}_C$ (0.60)	${}^5\chi^v{}_{CH}$ (0.67)
${}^0\chi$ (0.93)	CIC_2 (−0.87)	${}^5\chi^v{}_C$ (0.60)	${}^6\chi^b{}_{CH}$ (0.47)

The symbols and definitions of the indices shown in this Table can be found in Table 10.1. The bonding connectivity indices were defined for the first time by Basak et al. [19]

Some of the TIs used in this study, e.g., Randic's [60] first-order connectivity index (${}^1\chi$) and the information theoretic indices developed by Bonchev and Trinajstić [61] and Raychaudhury et al. [24], were used to discriminate the set of congeneric structures including alkanes. In the case of 18 octanes, the molecules do not vary much from one another with respect to size, but primarily in terms of branching patterns. Therefore, these indices were rightly interpreted based on those data as reflecting molecular branching. But when PCA was carried out with a diverse set of 3692 chemical structures, the *results entered an uncharted territory and were counterintuitive, to say the least.* As shown from the correlation of the original variables with PC_1, ${}^1\chi$ and related indices were now strongly correlated with molecular size in the large and diverse set, not to molecular branching. PC_3 emerged as the axis correlated with indices that encoded branching information, the cluster-type molecular connectivity indices in particular. *This result shows that the structural meaning of TIs that we derive intuitively or from correlational analyses is dependent on the nature and relative diversity of the structural landscape under investigation.* Further studies of TIs computed for both congeneric and diverse structures are needed to shed light on this important issue.

A virtual library of 248,832 psoralen derivatives [21] was created and analyzed using PCs derived from calculated TIs. *This set may be called congeneric because although it is a large collection of structures, it is derived from the same basic molecular skeleton: psoralen.* For this study, 92 topological indices were calculated by POLLY. In this set, the top 3 PCs explained 89.2 % of the variance in the data; first 6 PCs explained 95.5 % of the variance of the originally calculated indices. The PCs were used to cluster the large set of chemicals into a few smaller subsets as an exercise of managing *combinatorial explosion* that can happen in the drug design scenario when one wants to create a large pool of derivatives of a lead compound. For details of the outcome of clustering of the 248,832 psoralen derivatives, please see [21].

To conclude this section on the exploration of intrinsic dimensionality of structural spaces using PCA and calculated chemodescriptors, the data on the congeneric set of psoralens and the diverse set of 3, 692 TSCA chemicals appear to indicate that as compared to congeneric collections of structures, diverse sets need a higher number of orthogonal descriptors (dimensions) to explain a comparable amount of variance in the data. The fact that PCA brings down the number of descriptors from 90 or 92 calculated indices to 10 or 6 PCs keeping the explained variance at above 90 % level reflects that the intrinsic dimensionality of the structure space is adequately reflected by a small number of orthogonal variables. Thinking in terms of the philosophical idea known as the **Ockham's razor or the parsimony principle – it is futile to do with more what can be done with fewer** – PCA helps us to select a *useful and smaller subset of factors from a collection of many more*. To quote Hoffmann et al. [62]:

> Identifying the number of significant components enables one to determine the number of real sources of variation within the data. The most important applications of PCA are those related to: (a) classification of objects into groups by quantifying their similarity on the basis of the Principal Component scores; (b) interpretation of observables in terms of Principal Components or their combination; (c) prediction of properties for unknown samples. These are exactly the objectives pursued by any logical analysis, and the Principal Components may be thought of as the true independent variables or distinct hypotheses.

It is noteworthy that Katritzky et al. used PCA for the characterization of aromaticity [63] and formulation of QSARs [64] in line with the parsimony principle.

10.3.3 Some Examples of Hierarchical QSAR (HiQSAR) Using Calculated Chemodescriptors

10.3.3.1 Aryl Hydrocarbon (Ah) Receptor Binding Affinity of Dibenzofurans

Dibenzofurans are widespread environmental contaminants that are produced mainly as undesirable by-products in natural and industrial processes. The toxic effects of these compounds are thought to be mediated through binding to the aryl hydrocarbon (*Ah*) receptor. We developed HiQSAR models based on a set of 32 dibenzofurans with *Ah* receptor binding affinity values obtained from the literature [65]. Descriptor classes used to develop the models included the TS, TC, 3D, and the STO-3G class of *ab initio* QC descriptors. Statistical metrics for the ridge regression (RR), partial least square (PLS), and principal component regression (PCR) models are provided in Table 10.4. We found that the RR models were superior to those developed using either PLS or PCR. Examining the RR metrics, it is evident that the TC and the TS + TC descriptors provide high-quality predictive models, with R^2_{cv} values of 0.820 and 0.852, respectively. The addition of the 3D and STO-3G descriptors does not result in significant improvement in model quality. When each of these classes viz., 3-D and STO-3G quantum chemical descriptors, is used alone, the results are quite poor. This indicates that the topological indices are capable of adequately representing those structural features which are relevant to the binding of dibenzofu-

Table 10.4 Summary statistics for predictive *Ah* receptor binding affinity models

	$R^2_{c.v.}$			PRESS		
Independent variables	RR	PCR	PLS	RR	PCR	PLS
TS	0.731	0.690	0.701	16.9	19.4	18.7
TS+TC	0.852	0.683	0.836	9.27	19.9	10.3
TS+TC+3D	0.852	0.683	0.837	9.27	19.9	10.2
TS+TC+ 3D + STO-3G	0.862	0.595	0.862	8.62	25.4	8.67
TS	0.731	0.690	0.701	16.9	19.4	18.7
TC	0.820	0.694	0.749	11.3	19.1	15.7
3D	0.508	0.523	0.419	30.8	29.9	36.4
STO-3G	0.544	0.458	0.501	28.6	33.9	31.3

rans to the *Ah* receptor. Comparison of the experimentally determined binding affinity values and those predicted using the TS + TC RR model is available in Table 10.5. The details of this QSAR analysis has been published [66].

10.3.3.2 HiQSAR Modeling of a Diverse Set of 508 Chemical Mutagens

TS, TC, 3D, and QC descriptors for 508 chemical were calculated, and QSARs were formulated hierarchically using these four types of descriptors. For details of calculations and model building, see [67]. The method interrelated two-way clustering, ITC [56], which falls in the unsupervised class of approaches [68], was used for variable selection. Table 10.6 gives results of ridge regression (RR) alone as well as those where RR was used on descriptors selected by ITC. For both RR only and ITC+ RR analysis, the TS + TC combination gave the best models for predicting mutagenicity of the 508 diverse chemicals. The addition of 3-D and QC descriptors to the set of independent variables made minimum or no improvement in model quality.

Recent review of results of HiQSARs carried out by Basak and coworkers [46, 69–71] using topostructural, topochemical, 3-D, and quantum chemical indices for diverse properties, e. g., acute toxicity of benzene derivatives, dermal penetration of polycyclic aromatic hydrocarbons (PAHs), mutagenicity of a congeneric set of amines (heteroaromatic and aromatic), and others, indicates that in most of the above mentioned cases, TS+ TC combination of indices gives reasonable predictive models. The addition of 3-D and quantum chemical indices after the use of TS and TC descriptors did very little improvement in model quality.

How do we explain the above trend in HiQSAR? *One plausible explanation is that for the recognition of a receptor*, e.g., *the interaction of dibenzofuran with Ah receptor, discussed in* Sect. 10.3.3.1, *the dibenzofuran derivatives probably need some specific geometrical and stereo-electronic factors or a specific pharmacophore. But once the minimal requirement of this recognition is present in the molecule, the alterations in bioactivities from one derivative to another in the same structural class are governed by more general structural features which are quantified reasonably well by the TS and TC indices derived from the conventional bonding topology of molecules and features like sigma bond, π bond, lone pair of electrons, hydrogen bond donor acidity, hydrogen bond acceptor basicity*, etc. More studies with different groups of molecules with diverse bioactivities are needed to validate or falsify this hypothesis in line with the falsifiability principle of Sir Karl Popper [72], a basic scientific paradigm in the philosophy of science which defines the inherent testability of any scientific hypothesis.

Table 10.5 Experimental and cross-validated predicted *Ah* receptor binding affinities, based on the TS + TC ridge regression model of Table 10.4

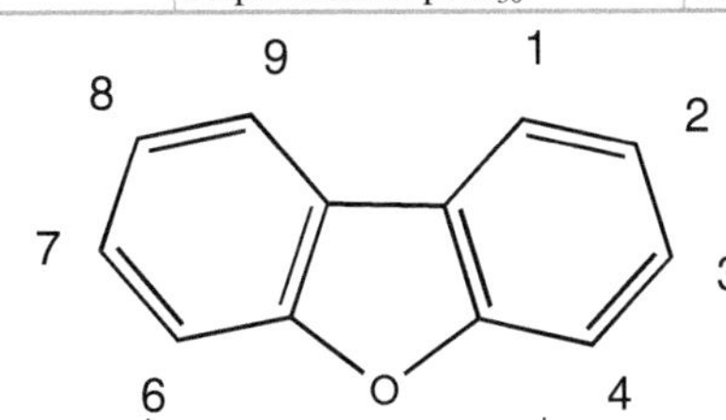

No.	Chemical	Experimental pEC_{50}	Predicted pEC_{50}	Exp. – Pred.
1	2-Cl	3.553	3.169	0.384
2	3-Cl	4.377	4.199	0.178
3	4-Cl	3.000	3.692	–0.692
4	2,3-diCl	5.326	4.964	0.362
5	2,6-diCl	3.609	4.279	–0.670
6	2,8-diCl	3.590	4.251	–0.661
7	1,2,7-trCl	6.347	5.646	0.701
8	1,3,6-trCl	5.357	4.705	0.652
9	1,3,8-trCl	4.071	5.330	–1.259
10	2,3,8-trCl	6.000	6.394	–0.394
11	1,2,3,6-teCl	6.456	6.480	–0.024
12	1,2,3,7-teCl	6.959	7.066	–0.107
13	1,2,4,8-teCl	5.000	4.715	0.285
14	2,3,4,6-teCl	6.456	7.321	–0.865
15	2,3,4,7-teCl	7.602	7.496	0.106
16	2,3,4,8-teCl	6.699	6.976	–0.277
17	2,3,6,8-teCl	6.658	6.008	0.650
18	2,3,7,8-teCl	7.387	7.139	0.248
19	1,2,3,4,8-peCl	6.921	6.293	0.628
20	1,2,3,7,8-peCl	7.128	7.213	–0.085
21	1,2,3,7,9-peCl	6.398	5.724	0.674
22	1,2,4,6,7-peCl	7.169	6.135	1.035
23	1,2,4,7,8-peCl	5.886	6.607	–0.720
24	1,2,4,7,9-peCl	4.699	4.937	–0.238
25	1,3,4,7,8-peCl	6.699	6.513	0.186
26	2,3,4,7,8-peCl	7.824	7.479	0.345
27	2,3,4,7,9-peCl	6.699	6.509	0.190
28	1,2,3,4,7,8-heCl	6.638	6.802	–0.164
29	1,2,3,6,7,8-heCl	6.569	7.124	–0.555
30	1,2,4,6,7,8-heCl	5.081	5.672	–0.591
31	2,3,4,6,7,8-heCl	7.328	7.019	0.309
32	Dibenzofuran	3.000	2.765	0.235

Table 10.6 HiQSAR model (RR and ITC + RR) for a diverse set of 508 chemical mutagens. All four means the model used TS+TC+3D+QC descriptors

Model type	Predictor type	Predictor number	% Correct classification	Sensitivity	Specificity
RR	TS	103	53.14	52.34	53.97
	TS+TC	298	76.97	83.98	69.84
	All four	307	77.17	84.38	69.84
ITC	TS	103	66.34	73.83	58.73
	TS+TC	298	73.23	77.34	69.05
	TS+TC+3D	301	74.80	77.34	72.22
	All four	307	72.05	76.17	67.86

10.3.4 Two QSAR Paradigms: Congenericity Principle Versus Diversity Begets Diversity Principle Analyzed Using Computed Mathematical Chemodescriptors of Homogeneous and Diverse Sets of Chemical Mutagens

The age-old paradigm of quantitative structure-activity relationship (QSAR) is the *congenericity principle* which states that similar structures usually have similar properties. But these days, a lot of large and structurally diverse data sets of chemicals with the same experimental data (dependent variable) are available. Starting with the same classes of descriptors, we extracted the two subsets of statistically most significant predictors for the formulation of QSARs for two sets of chemicals: a homogeneous set of 95 amine mutagens and a diverse set of 508 structurally diverse mutagens. The predictors included calculated TS, TC, geometrical, and QC indices. Whereas for the homogeneous amines, a small group of only seven descriptors were found to be significant in model building, for the 508 diverse set 42 descriptors were found to be statistically significant [73]. This preliminary and empirical study supports the ***DIVERSITY BEGETS DIVERSITY*** principle of QSAR formulated for the first time by Basak [18].

10.3.5 Applicability Domain of QSAR Models

A very important issue in the development of a QSAR model is that of defining the applicability domain (AD) of the model. This is necessary for any valid implementable QSAR model according to OECD principles [74]. There are a few methods of defining the AD of statistical models which can be roughly divided into two classes: (a) AD methods that define the active predictor space through some method like bounding box, PCA, or convex hulls and (b) distance-based methods which compute the similarity/dissimilarity of a new compound to the set of compounds which have been used in formulating the training QSAR model. To obtain predictions for any incoming sample set using the model, the first group of methods is used to ensure that the compounds are within the so-called active subspace: which essentially means we are actually performing interpolation, not extrapolation [75, 76]. For the distance-based approach, a predefined statistic is calculated to quantify the proximity of the test compounds to the training set, and based on whether that statistic is above or below a certain cutoff value, predictions for that compound are considered reasonable or not [75, 77].

10.3.6 Practical Applications of QSAR

> Knowledge is of no value unless you put it into practice.
>
> Anton Chekhov

Practical applications of good quality QSARs, particularly those based on easily calculable molecular descriptors, can be very useful tools in pharmaceutical drug design and specialty chemical design.

The journey of identified lead molecules in the drug discovery pipeline is a long and risky one. Average cost of developing a drug (including the cost of failures) during 2000s to early 2010s was US \$2.6 billion [78]. One important contributing factor to this astronomical cost is that the drug developer has to produce and test a large number of derivatives of the lead structure for their beneficial and toxic side effects before one marketable drug is found. QSAR plays a very important role in drug design providing a cheaper and fast alternative to the medium throughput *in vitro* and low throughput *in vivo* screening of chemicals, which are generally used more frequently in the later stages of the discovery cascade. It has been noted that currently no drug is developed without going through the prior evaluation by QSAR methods [79].

In Fig. 10.4, a generic scheme is presented for the use of QSAR in drug discovery. Starting with a "lead," modern combinatorial chemistry can produce millions, even billions, of derivatives. Such real or hypothetical chemicals must be evaluated in real time to prioritize them for synthesis and testing. QSARs based on easily calculated descriptors can help us in accomplishing this task.

The era of "Big Data" has arrived in the realm of drug discovery. For a concise description of trends in this realm, please see Basak et al. [80].

10.4 Molecular Similarity and Tailored Similarity Methods

> Like substances react similarly and similar changes in structure produce similar changes in reactivity
>
> L P. Hammett

> All cases are unique, and very similar to the others.
>
> *T.S. Eliot*, In: The Cocktail Party

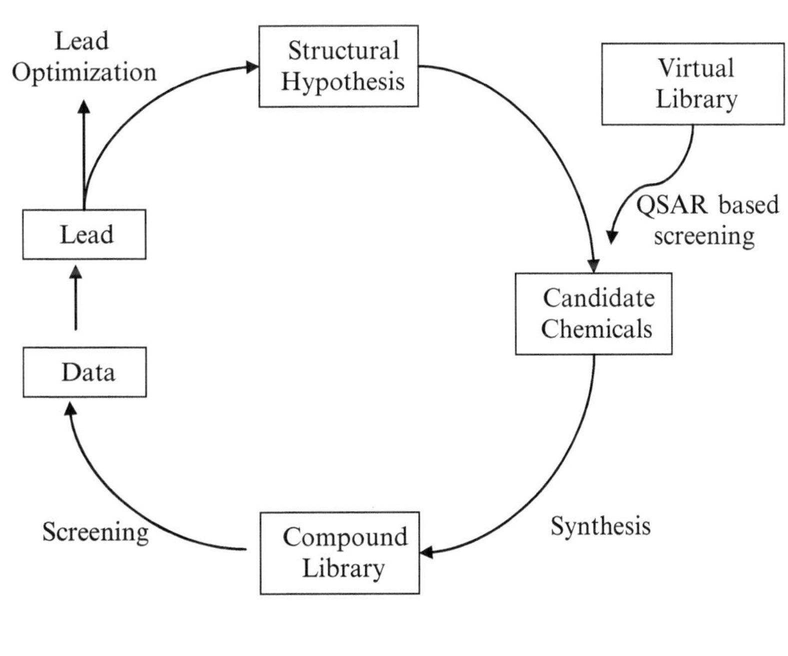

Fig. 10.4 A generic scheme for the use of QSARs in drug discovery protocols

Molecular similarity is a well-known concept, which is intuitively understood by many researchers. There is a tacit consensus among molecular similarity researchers that *similar structures usually have similar properties*. In a broader scope, this "*structure-property similarity principle*" includes the notion that similar "structural organizations" of objects lead to similar observable properties. In the realms of chemistry, biology, and toxicology, the natural extension of this structure-property similarity principle is that atoms, ions, molecules, and macromolecules with similar structures will have similar physicochemical, biological, and toxicological properties. This principle is vindicated by a vast majority of facts at varying levels of structural organization.

In the realm of cellular biochemistry, the inhibition of succinic dehydrogenase by malonate *in vitro* is explained in terms of the competition by malonate for the active sites of the enzyme succinic dehydrogenase, arising from the structural similarity between the substrate succinic acid and malonic acid [81, 82]. This is probably one of the earliest observations of the inhibition of an enzyme by an analog of its substrate. Another well-known example is that the structural similarities between p-amino benzoic acid and sulfanilic acid allow both compounds to interact with a specific bacterial biosynthetic enzyme. This "case of mistaken identity" is the basis for the antibacterial activity of sulfonamide antimicrobials [1].

There is no consensus regarding the optimal quantification of molecular similarity. In most cases, measures of molecular similarity are defined by the individual practitioner, generally based on his/her experience in a particular research area or some intuitive notion. If the researcher selects n different attributes for the molecules under investigation, then the molecules can be looked upon as points in some type of n-dimensional space. A distance function can then be used to measure the distance between various objects (chemicals) in that space, and the magnitude of distance serves as a measure of the degree of similarity or dissimilarity between any pair of molecules in this n-dimensional similarity space. Difficulties arise from two major factors: (1) the selection of appropriate axes for developing the similarity space and (2) the relevance of the selected axes to the property under investigation. Many molecular similarity scientists have their own favorite measures, but the axes selected might be multicollinear or may encode essentially the same information multiple times. One popular solution for this problem is the use of orthogonal axes derived from the original axes using techniques such as PCA mentioned above. A more serious concern is whether or not the subjectively chosen axes are relevant to the property under investigation. This is a more difficult problem to address. One potential solution to this issue, pursued by our research group, is the use of the *tailored similarity* method (*vide infra*).

One practical application of molecular similarity in pharmaceutical drug design, human health hazard assessment, and environmental risk analysis is the selection of analogs. Once a lead structure with interesting properties is found, the drug designer often asks "*Is there a chemical similar in structure to the lead, which also has analogous properties*?" In contemporary drug discovery research, scientists usually search various proprietary and public domain databases for chemical analogs. Analogs can be selected based on the researcher's intuitive notion of chemical similarity, their similarity with respect to measured properties, or calculated molecular descriptors. Since most of the chemicals in many databases have very little available experimental property data, similarity methods based on calculated properties or molecular descriptors are used more frequently for analog selection. In environmental risk analysis, analogs of suspected toxicants or newly produced industrial chemicals are used in hazard assessment when the molecule is so unique or so complex that class-specific QSARs cannot be applied in toxicity estimation [36]. The flip side of similarity is dissimilarity. This concept can be applied to both drug discovery and predictive toxicology to reduce the number of compounds in the database from a combinatorial explosion to a manageable number that can be handled through laboratory testing. One such example was discussed above in Sect. 10.3.2 for the case of a large virtual library of 248,832 psoralen derivatives which were clustered using PCs extracted from 92 computed POLLY indices.

10.4.1 Arbitrary or User-Defined Similarity Methods

In *arbitrary similarity methods*, one subjectively defines the similarity measure. In essence, the experienced practitioner says "*My personal experience with data or my intuitive notion tells me that the prescribed similarity measures will lead to useful grouping of chemicals with respect to the property of interest.*" This might work out in narrowly defined cases, but in complex situations where a large number of parameters are needed to characterize the property, intuition is usually less accurate. Also, one may want to select analogs which are ordered with respect to widely different properties of the same chemical, e.g., carcinogenicity versus boiling point. The same intuitive measure cannot give "good analogs" for properties that are not mutually correlated. Various authors have used apparently diverse, arbitrary similarity measures in an effort to select mutually dissimilar analogs, but the rational basis of such selections has never been clear. The tailored approach to molecular similarity may help solve this issue.

10.4.1.1 Probing the Utility of Five Different Similarity Spaces

A wide variety of chemical information can and have been used in developing molecular similarity spaces. Many researchers contend that similarity spaces derived from physicochemical property data are inherently better, since the results are much more readily interpretable. However, as was stated earlier, physicochemical property data is not widely available for many chemicals, thus necessitating the use of calculated descriptors. One interesting aspect of research in the field of molecular similarity has been the comparison of arbitrary similarity spaces derived from physicochemical properties with spaces derived from calculated molecular descriptors. For a recent review on the topic of quantitative molecular similarity analysis studies carried out by Basak and coworkers, please see [22].

In a 1995 study, Basak and Grunwald [83] developed five distinct similarity spaces and tested those on a set of 73 aromatic and heteroaromatic amines with known mutagenicity (ln Rev/nmol) data. The derived similarity spaces were based on quantum theoretical descriptors believed to correlate well with mutagenicity (property), principal components derived from those descriptors (PC_{Prop}), atom pairs (APs), principal components derived from a set of topological indices (PC_{TI}), and principal components derived from the combined set of quantum theoretical descriptors and topological indices (PC_{All}). While the similarity spaces derived from the quantum theoretical descriptors resulted in the best correlations with mutagenicity, spaces derived from atom pairs and the combined set of topological and quantum theoretical descriptors estimated mutagenicity nearly as well. The results for the five similarity spaces are summarized in Table 10.7, where r is the correlation coefficient, $s.e.$ is the standard error, n is the number of dimensions or axes in the similarity space, and k is the number of selected "nearest neighbors" used to estimate mutagenicity for each chemical within the space.

10.4.1.2 Molecular Similarity and Analog Selection

As mentioned earlier, many times a researcher's goal is to select a set of analogs for a chemical of interest from a large, diverse data set based on similarity spaces derived solely from calculated

Table 10.7 Comparison of five similarity methods in the estimation of mutagenicity (ln Rev/nmol in *S. typhimurium* TA100 with metabolic activation) for 73 aromatic and heteroaromatic amines

Similarity method	r	$s.e.$	n	k
AP	0.77	0.88	na	4
PC_{TI}	0.72	0.96	6	5
Property	0.83	0.77	3	5
PC_{Prop}	0.84	0.75	3	5
PC_{All}	0.79	0.85	7	4

descriptors of molecular structure. We described above in Sect. 10.3.2 our PCA analysis of the diverse set of 3692 industrial chemicals [19]. As part of this study, analogs were selected based on *Euclidean distance* within the ten-dimensional similarity space derived from the ten major principal components. Figure 10.5 presents an example of the five nearest neighbors (or analogs) selected for one chemical from the set of 3692 molecules.

A look at the five selected structures, particularly the ones closest to 4-hydroxybenzene acetic acid (the probe or query chemical), shows that there is sufficient degree of similarity of the query structure with the selected analogs in terms of the number and type of atoms, degree of cyclicity, aromaticity, etc.

10.4.1.3 The K-Nearest Neighbor (KNN) Approach in Predicting Modes of Action (MOAs) of Industrial Pollutants

Different domains of chemical screening use different model organisms for the assessment of bioactivity of chemicals. In aquatic toxicology and ecotoxicology, *fathead minnow* is an important model organism [84–86]. Numerous QSARs have been developed with subsets of fathead minnow toxicity (LC_{50}) data, many such models being developed using small, structurally related or congeneric sets. But, following the *diversity begets diversity principle* discussed above, one will need a diverse collection of molecular descriptors for the QSAR formulation of diverse collection of chemicals. Another possibility is to develop different subsets of chemicals from a large and diverse set based on their *mode of action* (MOA) first and then treat chemicals with the same MOA as *biological congeners* as opposed to structural classes which may be called *structural congeners*. Basak et al. [87] undertook a classification study based on acute toxic MOA of industrial chemicals. At that time the US Environmental Protection Agency's Mid-Continent Ecology Division-Duluth, Minnesota, fathead minnow database had LC_{50} data on 617 chemicals. But out of that list, only 283 chemicals were selected by us because our experimental cooperators had good confidence about the MOAs of that subset only. Such evidence consisted of concurrent information from joint chemical toxicity studies, physicochemical and behavioral response, information published in peer-reviewed literature, and toxicity over time [88]. Such caution in the selection of good subsets of data for modeling is in line with the *veracity attribute* mentioned above while discussing the major pillars of QSAR and issues regarding Big Data [80].

Acute toxic mode of action of the chemicals was predicted using molecular similarity method, neural networks of the Learning Vector Quantization (LVQ) type, and discriminant analysis methods. The set of 283 compounds was broken down into

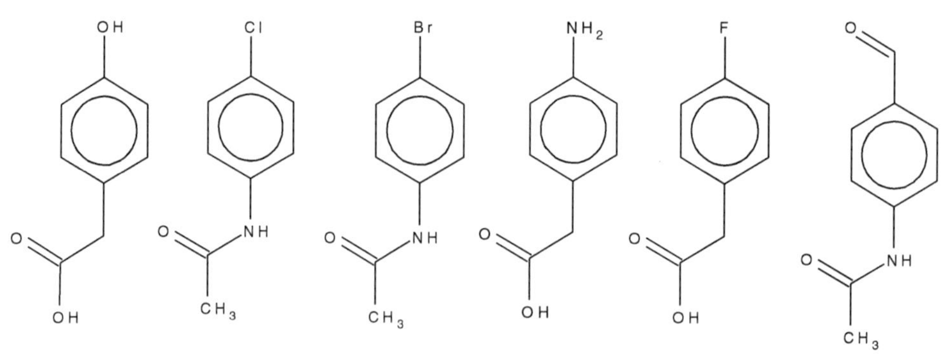

ED	0.230	0.486	0.488	0.520	0.584

Fig. 10.5 Molecular structures for 4-hydroxybenzeneacetic acid and its five analogs selected from a database of 3692 chemicals. The numbers below each structure are the Euclidean distances (*ED*) between 4-hydreoxybenzeneacetic acid (the left-most structure) and its analogs

a training set of 220 compounds and a test set of 63. Computed topological indices and atom pairs were used as structural descriptors for model development. The five MOA classes represented included:

1. Narcosis I/II and electrophile/proelectrophile reactivity (NE)
2. Uncouplers of oxidative phosphorylation (UNC)
3. Acetylcholinesterase inhibitor (AChE-I)
4. Neruotoxicants (NT)
5. Neuordepressants/respiratory blockers (RB/ND)

In the molecular similarity approach, similarity between chemicals i and j was defined as

$$S_{\mathrm{ij}} = 2C / \left(T_{\mathrm{i}} + T_{\mathrm{j}}\right) \tag{10.4}$$

where C is the number of atom pairs [10] common to molecules i and j. $T_{\mathrm{i}} + T_{\mathrm{j}}$ are the total number of atom pairs in i and j, respectively. The five nearest neighbors (i.e., $K=5$) were used to predict the mode of action of a probe or query chemical.

In the neural network analysis, LVQ classification network was used, consisting of a 60-node input layer, a 5-node hidden layer, and a 5-node output layer.

Linear models utilizing stepwise discriminant analysis were developed in addition to the neural network and similarity models.

All three methods gave good results for training and test sets, with the success ranging from 95 % for the K-nearest neighbor method to 87 % for the discriminant analysis technique. This consistency of results obtained using topological descriptors in different classification methods indicates that the graph theoretical parameters used in this study contain sufficient structural information to be capable of predicting modes of action of diverse chemical species. Table 10.8 provides the classification results obtained using the K-nearest neighbor method, in which 90 % of the training set chemicals and 95 % of the test set chemicals were classified correctly.

10.4.1.4 The *Tailored Approach* to Developing Similarity Spaces

From the words of the poet, men take what meanings please them; yet their last meaning points to thee.

Rabindranath Tagore, Poem #75
Gitanjali

As mentioned above, user-defined or arbitrary molecular similarity methods perform reasonably well in narrow, well-defined situations. But the relationship between structural attributes and biomedicinal or toxicological properties are not always crisp; they are often messy. Human intuition often fails in such circumstances. Similarity methods based on objectively defined relationships are needed, rather than those derived from subjective or intuitive approaches. In a multivariate space, this should be accomplished using robust statistical methods. The *tailored similarity method* starts with an appropriate number of molecular descriptors [89–91]. These descriptors are run through *ridge regression* analysis modeling the property of interest, and a small number of independent variables with high |*t*| values are selected as the axes of the similarity space. In this way, we select variables which are strongly

Table 10.8 MOA classification results using the K-nearest neighbor ($K=5$) method

	Training set $n=220$	% Correct
NE	180/183	98 %
UNC	6/10	60 %
AChE-I	7/14	50 %
NT	0/7	0 %
RB/ND	5/6	83 %
	Overall	**90 %**

	Test set $n=63$	% Correct
NE	53/54	98 %
UNC	2/2	100 %
AChE-I	3/3	100 %
NT	1/2	50 %
RB/ND	1/2	50 %
	Overall	**95 %**

related with the property of interest instead of a subjectively selected group of descriptors. Needless to say, human intuition will be hard pressed to match the objective relationship developed by ridge regression techniques.

In one tailored similarity study [91], we examined the effects of tailoring on the estimation of logP for a set of 213 chemicals and on the estimation of mutagenicity for a set of 95 aromatic and heteroaromatic amines. In this study we utilized a much larger set of topological indices than have been used in many of our *earlier* studies. Three distinct similarity spaces were constructed, though two were "overlapping" spaces. The overlapping spaces were derived using principal component analysis on the set of 267 topological indices. The PCA created 20 orthogonal components with eigenvalues greater than one. These 20 PCs were used as the axes for the first similarity space. The second similarity space was derived from the principal components. In examining the PCs, we selected the index most correlated with each cluster as a representative of the cluster. One of the arguments against using PCA to reduce the number of variables for modeling is that PCs, being linear combinations of the indices, are not easily interpretable. So, by selecting the most correlated single TI from each PC, we have a set of easily interpretable topological indices to use in modeling.

Finally, the third set of indices was selected based on a ridge regression model developed from all 267 indices to predict mutagenicity. From the modeling results, *t*-values were extracted and the 20 indices with the highest absolute [t] values were selected as axes for developing the similarity space. A summary of the correlation coefficients for estimating mutagenicity from the three similarity spaces for varying numbers of neighbors using the KNN method is presented in Fig. 10.6.

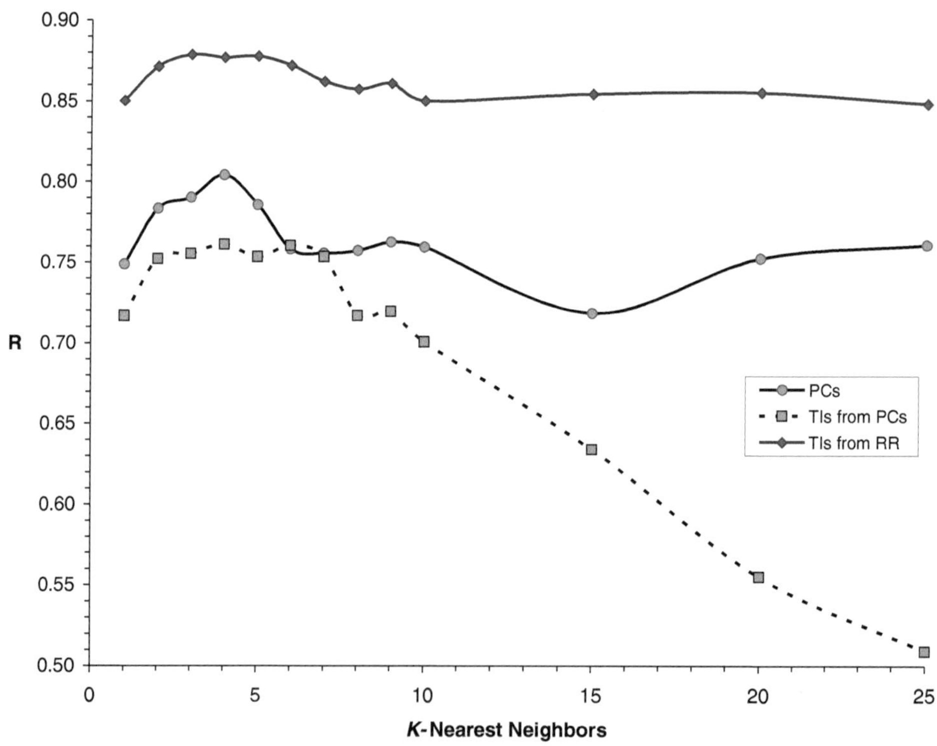

Fig. 10.6 Plot of the pattern of correlation coefficient (R) from $k=1$–10, 15, 20, and 25 for the estimation of mutagenicity (ln Rev/nmol) for 95 aromatic and heteroaromatic amines using a 20 principal component space derived from 267 topological indices (PCs), a 20 topological index space selected from the principal components (TIs from PCs), and a 20 topological index based on space derived from ridge regression (TIs from RR)

It is clear from Fig. 10.6 that tailoring the selected set of indices significantly improved the estimative power of the model, resulting in roughly a 10 % increase to the correlation coefficient. These results, as with all of the results we have seen from tailored similarity spaces, are promising, and we believe that **tailored similarity methods will be very useful both in drug discovery and toxicological research**.

10.5 Formulation of Biodescriptors from DNA/RNA Sequences and Proteomics Maps: Development and Applications

If your chromosomes are XYY,
And you are a naughty, naughty guy,
Your crimes, the judge won't even try,
'Cause you have a legal reason why
He'll raise his hands and gently sigh!
"I guess for this you get a by."
By Carl A. Dragstedt
In: Perspectives in Biology and Medicine
Vol. 14, # 1, autumn, 1970

10.5.1 Mathematical Biodescriptors from DNA/RNA Sequences

After the completion of the *Human Genome Project*, a lot of data for DNA, RNA, and protein sequences are being generated. In line with the idea of representation and mathematical characterization of chemicals (see Fig. 10.2 above), various authors have developed such representation-cum-characterization methods for DNA/RNA sequences [16, 92–96]. In the past few years, a lot of papers have been published in this area. Here, we give a brief history of the recent growth spurt of this exciting field beginning in 1998. Dilip K. Sinha and Subhash C. Basak started the Indo-US Workshop Series on Mathematical Chemistry [97] in 1998, the first event being held at the Visva Bharati University, Santiniketan, West Bengal, India. Raychaudhury and Nandy [98] gave a presentation on mathematical characterization of DNA sequences using their graphical method. This caught the attention of Basak who later developed a research group on the mathematical characterization of DNA/RNA sequences supported by funds from the University of Minnesota Duluth-Natural Resources Research Institute (UMD-NRRI) and University of Minnesota. This led to the publication of the first couple of papers on DNA sequence invariants [99, 100]. The rest of the development of DNA/RNA sequence graph invariants and mathematical descriptors is clear from the hundreds of papers published on this topic subsequently by authors all over the world. More recently Nandy and Basak applied this method in the characterization of the various bird flu sequences, e.g., *H5N1 bird flu* [101] and *H5N2 pandemic bird flu* [102], the latter one causing havoc in the turkey and poultry farms of the Midwest of the USA in 2015. Numerous other theoretical developments and practical applications of DNA/RNA mathematical descriptors are not discussed here for brevity.

10.5.2 Mathematical Proteomics-Based Biodescriptors

Proteomics may be looked upon as a branch of Functional Genomics that studies changes in protein-protein and protein-drug/toxicant interactions. Scientists are studying proteomics for new drug discovery and predictive toxicology [103–105]. A typical 2D gel electrophoresis (2DE)-derived proteomics map provided to us by our collaborators at Indiana University is provided in Fig. 10.7.

The 2DE method of proteomics is capable of detecting and characterizing a few thousand proteins from a cell, tissue, or animal. One can then study the effects of well-designed structural or mechanistic classes of chemicals on animals or specialized cells and use these proteomics data to classify the molecules or predict their biological action. But with 1000–2000 protein spots present per gel, the difficult question we face is: **How do we make sense of the chaotic pattern of the large number of proteins as shown in Fig. 10.7?**

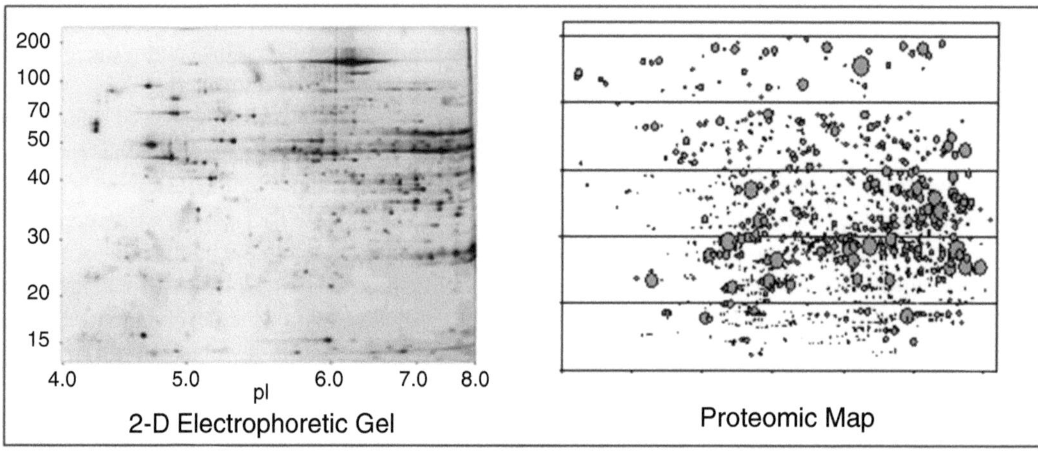

Fig. 10.7 Location and abundance of protein spots derived from 2D gel electrophoresis (Courtesy of Frank Witzmann of Indiana University, Indianapolis, USA)

We have attacked this problem through the formulation of biodescriptors applying the techniques of *discrete mathematics* to proteomics maps. Described below are three major approaches developed by our research team at the Natural Resources Research Institute and its collaborators for the quantitative calculation of biodescriptors of proteomics maps, the term **biodescriptor** being coined by the Basak group for the first time:

(a) In each 2D gel, the proteins are separated by charge and mass. Also associated with each protein spot is a value representing abundance, which quantifies the amount of that particular protein or closely related class of proteins gathered on one spot. Mathematically, the data generated by 2DE may be looked upon as points in a three-dimensional space, with the axes described by charge, mass, and spot abundance. One can then have projections of the data to the three planes, i.e., XY, YZ, and XZ. The *spectrum-like data* so derived can be converted into vectors, and similarity of proteomics maps can be computed from these map descriptors [106].
(b) In a second approach, viz., the graph invariant biodescriptor method, different types of embedded graphs, e.g., zigzag graphs neibhborhood graphs, are associated with proteomics maps, with the set of spots in the proteomics maps representing the vertices of such graphs. In the zigzag approach, one begins with the spot of the highest abundance and draws an edge between it and the spot having the next highest abundance and continues this process. The resulting zigzag curve is converted into a *D/D matrix* where the (i, j) entry of such a matrix is the quotient of the Euclidean distance and the through-bond distance. For details on this approach, please see [107].
(c) A proteomics map may be looked upon as a pattern of protein mass distributed over a 2D space. The distribution may vary depending on the functional state of the cell under various developmental and pathological conditions as well as under the influence of exogenous chemicals such as drugs and xenobiotics. Information theoretic approach has been applied to compute biodescriptors called *map information content* (*MIC*) from 2D gels [108].

10.6 Combined Use of Chemodescriptors and Biodescriptors for Bioactivity Prediction

We told above in Eq. 10.2 that in many cases, the property/bioactivity/toxicity of chemicals can be predicted reasonably well using their structure (S) alone. But in many complex biological situations, e.g., induction of cancer by exposure to chemical carcinogens, we need to use both structural features of such chemicals and biological test data to make sense of such endpoints. Arcos [109], for example, suggested the use of specific biological data, e.g., degranulation of endoplasmic reticulum, peroxisome proliferation, unscheduled DNA synthesis, antispermatogenic activity, etc., as biological indicators of carcinogenesis. Such biochemical data not only bring direct and relevant biological observations into the set of predictors, they also bring independent variables which are closer to the endpoint in the scale of complexity than the chemical structure. In line with this *structural*-cum-*functional approach* in predicting bioactivity of chemicals, we have used a combination of chemodescriptors and proteomics-based biodescriptors for assessing toxicity of priority pollutants [28, 110].

10.7 Discussion

> We are all agreed that your theory is crazy. The question which divides us is whether it is crazy enough to have a chance of being correct. My own feeling is that it is not crazy enough.
>
> Niels Bohr

> Everything should be made as simple as possible, but not simpler.
> – Albert Einstein

Major objectives of this chapter have been to review our research in the use of mathematical chemodescriptors and biodescriptors in the prediction of bioactivity/toxicity of chemicals, quantification of similarity/dissimilarity among chemical species from their chemodescriptors, and similarity-based clustering, as well as estimation of toxicologically relevant properties of diverse groups of molecules.

In the chemodescriptor area, our major goal has been to review the utility of graph theoretical parameters, also known as topological indices, in QSAR and QMSA studies. We studied the intercorrelation of major topological indices in an effort to identify subsets that are minimally correlated [57, 111]. We have also used principal components derived from TIs and all TIs simultaneously (e.g., ridge regression models) in QSAR formulation. At present a large number of descriptors can be calculated for chemicals using available software. If the number of experimental data points (dependent variables) for QSAR model building is much smaller than the number of descriptors, i.e., the situation is rank-deficient, one needs to be cautious. We have discussed the variable selection methods including ITC [56] which, to our knowledge, has been brought to QSAR from the genomics/ genetics area for the first time in our research. In the calculation of q^2 in the rank-deficient case, one must follow the *two-deep cross-validation* procedure; otherwise the calculated q^2 will reflect overfitting [43–45, 51, 52, 55]. We have demonstrated this using one example where we deliberately used the wrong ordinary least square (OLS) approach in a rank-deficient case and compared the results with the correct approach to show the difference between them [45]. In HiQSAR modeling, we found that of the four types of calculated molecular descriptors, viz., TS, TC, 3-D, and QC indices, in the majority of cases a TS + TC combination gave good quality models; the addition of 3-D or QC descriptors after the use of TS and TC combination did not improve much the model quality. This is a good news in view of the fact that we are already at the age of *Big Data* [80] and easily calculated indices like TS and TC descriptors, if they give good models in many areas, could find wide applications in the *in silico* screening of chemicals. The *congenericity principle* has been a major theme of QSAR whereby there has been a tendency in developing QSARs of congeneric sets of chemicals. When the same property, viz., mutagenicity, of congeneric versus diverse sets was used to develop QSAR models, the congeneric set of 95 amines had much lower number of significant descriptors as compared to the diverse set of 508 molecules. This gives support to the *diversity begets diversity principle* formulated by us [18].

When a large number of descriptors are calculated for a set of chemicals, the data set becomes high dimensional. The use of PCA can derive a much smaller number of orthogonal variables which reflect the *parsimony principle* or *Occam's razor* [62].

Molecular similarity is used both in drug design and hazard assessment of chemicals [36,

39, 112]. We used calculated TIs and atom pairs to generate similarity spaces following different methods and used both Euclidean distance derived from PCs and Tanimoto coefficient based on atom pairs to select analogs. The structures of analogs selected from the structurally diverse set of 3692 industrial chemicals indicated that the calculated property-based QMSA methods are capable of selecting analogs of query chemicals that look reasonably structurally similar to them. We also used our QMSA method in selecting analogs of environmental pollutants for which the modes of action are known with high confidence from experimental toxicology. The results of the MOA prediction study show that selected analogs of chemicals with specified MOA fall in similar toxicological categories.

In the post-*genomic era*, the *omics* technologies are generating a lot of data on the effects of chemicals on the genetic system, viz., transcription, translation, and posttranslational modification, of the cell and tissue. We have been involved in the development of biodescriptors from DNA/RNA sequences and two-dimensional gel electrophoresis (2DE) data derived from cells/tissue exposed to drugs and toxicants. Results of our research in this area show that the biodescriptors developed from proteomics maps are capable of characterizing the pharmacological/toxicological profiles of chemicals [106–108]. Some preliminary studies have been done on the use of the combined set of chemodescriptors and biodescriptors in predicting bioactivity. Further research are needed to test the relative effectiveness of the two classes of descriptors, chemodescriptors versus biodescriptors, in predictive pharmacology and toxicology [28, 110].

At this juncture, after reviewing results of a large number of QSAR studies using chemodescriptors and biodescriptors, we may ask ourselves: *Quo Vadimus*? We have seen that calculated chemodescriptors are capable of predicting and characterizing bioactivity and toxicity as well as toxic modes of action of chemicals. Research using biodescriptors of different types also shows that such descriptors derived from proteomics maps have reasonable power of discriminating among structurally closely related toxicants. Can we, at this stage, opt for either chemodescriptor or biodescriptors alone? The answer is *no*, as is evident from our experience in predictive toxicology. This indicates that in the foreseeable future, we will need an integrated approach consisting of chemodescriptors and biodescriptors in order to obtain the best results (Fig. 10.8).

As discussed by this author [113] in a recent book on Advances in Mathematical Chemistry and applications:

> Mathematical chemistry or more accurately discrete mathematical chemistry had a tremendous growth spurt in the second half of the twentieth century and the same trend is continuing now. This growth was fueled primarily by two major factors: (1) Novel applications of discrete mathematical concepts to chemical and biological systems, and (2) Availability of high speed computers and associated software whereby *hypothesis driven* as well as *discovery oriented* research on large data sets could be carried out in a timely manner. This led to

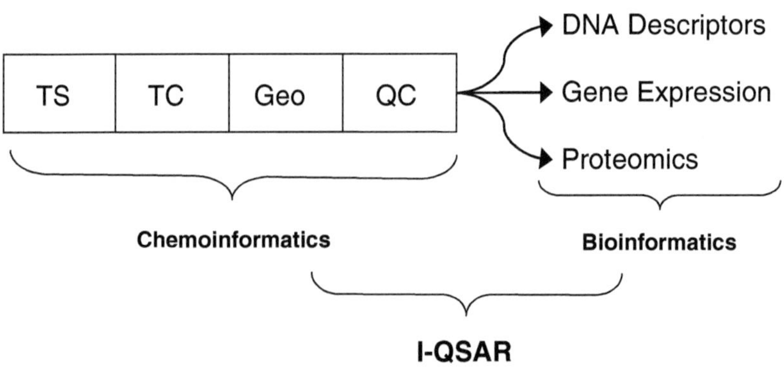

Fig. 10.8 Integrated QSAR, combining chemodescriptors and biodescriptors

the development of not only a plethora of new concepts, but also various useful applications to such important areas as drug discovery, protection of human as well as ecological health, bioinformatics, and chemoinformatics. Following the completion of the Human Genome Project in 2003, discrete mathematical methods were applied to the "omics" data to develop descriptors relevant to bioinformatics, toxicoinformatics, and computational biology.

The results of various types of research using chemodescriptors and biodescriptors [16–21, 28, 108, 114] derived through applications of discrete mathematics on chemical and biological systems give us hope that an exciting future is in front of us.

Acknowledgments I am thankful to Kanika Basak, Gregory Grunwald, Douglas Hawkins, Brian Gute, Subhabrata Majumdar, Denise Mills, Dilip K. Sinha, Ashesh Nandy, Frank Witzmann, Kevin Geiss, Krishnan Balasubramanian, Ramanathan Natarajan, Gerald J. Niemi, Alexandru T. Balaban, the late Alan Katritzky, Milan Randic, Nenad Trinajstic, Sonja Nikolic, Marjan Vracko, Marjana Novic, Xiaofeng Guo, Terry Neumann, Qianhong Zhu, late Gilman D. Veith, Marissa Harle, Vincent R. Magnuson, Donald K. Harriss, Chandan Raychaudhury, Samar K. Ray and Lester R. Drewes for collaboration in my research.

References

1. Hardman JG, Limbird LE, Gilman AG (2001) Goodman and Gilman's the pharmacological basis of therapeutics. McGraw- Hill, New York
2. Hoffman DJ, Ratner BA, Burton GA Jr, Cairns J Jr (1995) Handbook of ecotoxicology. CRC Press, Boca Raton
3. Nogrady T (1985) Medicinal chemistry: a biochemical approach. Oxford University Press, New York
4. Rand G (ed) (1995) Fundamentals of aquatic toxicology: effects, environmental fate and risk assessment, 2nd edn. Taylor and Francis, New York
5. Primas H (1981) Chemistry, quantum mechanics and reductionism. Springer, Berlin
6. Woolley RG (1978) Must a molecule have a shape? J Am Chem Soc 100:1073–1078
7. Basak SC, Veith GJ, Niemi GD (1991) Predicting properties of molecules using graph invariants. J Math Chem 7:243–272
8. Einstein A (1954) Remarks on Bertrand Russell's theory of knowledge. In: Einstein A (ed) Ideas and opinions. Ed. Carl Seelig, (Based on MEIN WELTBILD, edited by Carl Seelig, and other sources; New translations and revisions by Sonja Bargmann), Crown Publishers, New York, pp 18–24
9. Bunge M (1973) Method, model and matter. Reidel, Dordrecht
10. Carhart RE, Smith DH, Venkataraghavan R (1985) Atoms pairs as molecular features in structure-activity studies: definition and applications. J Chem Inf Comput Sci 25:64–73. doi:10.1021/ci00046a002
11. Euler I (1736) Solutio problematis ad geometriam situs pertinentis. Comment Acad Sci U Petrop 8:128–140
12. Sylvester JJ (1878) On an application of the new atomic theory to the graphical representation of the invariants and covariants of binary quantics, with three appendices. Am J Math 1:105–125
13. http://www.goodreads.com/quotes/926286-chemistry-has-the-same-quickening-and-suggestive-influence-upon-the
14. Wiener H (1947) Structural determination of paraffin boiling points. J Am Chem Soc 69:17–20
15. Balasubramanian K, Basak SC (1998) Characterization of isospectral graphs using graph invariants and derived orthogonal parameters. J Chem Inf Comput Sci 38:367–373
16. Nandy A, Harle M, Basak SC (2006) Mathematical descriptors of DNA sequences: development and application. Arkivoc 9:211–238
17. Basak SC (2013) Philosophy of mathematical chemistry: a personal perspective. HYLE Int J Philos Chem 19:3–17
18. Basak SC (2013) Mathematical descriptors for the prediction of property, bioactivity, and toxicity of chemicals from their structure: a chemical-cum-biochemical approach. Curr Comput Aided Drug Des 9:449–462
19. Basak SC, Magnuson VR, Niemi GJ, Regal RR (1988) Determining structural similarity of chemicals using graph-theoretic indices. Discrete Appl Math 19:17–44
20. Lajiness M (1990) Molecular similarity-based methods for selecting compounds for screening. In: Rouvray DH (ed) Computational chemical graph theory. Nova, New York, pp 299–316
21. Basak SC, Mills D, Gute BD, Balaban AT, Basak K, Grunwald GD (2010) Use of mathematical structural invariants in analyzing, combinatorial libraries: a case study with psoralen derivatives. Curr Comput Aided Drug Des 6:240–251
22. Basak SC (2014) Molecular similarity and hazard assessment of chemicals: a comparative study of arbitrary and tailored similarity spaces. J Eng Sci Manag Educ 7:178–184
23. Basak SC (1987) Use of molecular complexity indices in predictive pharmacology and toxicology: a QSAR approach. Med Sci Res 15:605–609
24. Raychaudhury C, Ray SK, Ghosh JJ, Roy AB, Basak SC (1984) Discrimination of isomeric structures

using information-theoretic topological indices. J Comput Chem 5:581–588
25. Balaban AT, Mills D, Ivanciuc O, Basak SC (2000) Reverse wiener indices. Croat Chim Acta 73:923–941
26. Nikolic S, Trinajstic N, Amic D, Beslo D, Basak SC (2001) Modeling the solubility of aliphatic alcohols in water. Graph connectivity indices versus line graph connectivity indices. In: Diudea MV (ed) QSAR/QSPR studies by molecular descriptors. Nova, Huntington, pp 63–81
27. Randic M, Vracko M, Nandy A, Basak SC (2000) On 3-D graphical representation of DNA primary sequences and their numerical characterization. J Chem Inf Comput Sci 40:1235–1244
28. Basak SC, Gute BD (2008) Mathematical descriptors of proteomics maps: background and applications. Curr Opin Drug Discov Dev 11:320–326
29. Hosoya H (1971) Topological index. A newly proposed quantity characterizing the topological nature of structural isomers of saturated hydrocarbons. Bull Chem Soc Jpn 44:2332–2339
30. MolconnZ (2003) Version 4.05. Hall Ass. Consult. Quincy
31. Basak SC, Harriss DK, Magnuson VR (1988) POLLY v. 2.3. Copyright of the University of Minnesota, USA
32. Basak SC, Grunwald GD (1993) APProbe. Copyright of the University of Minnesota, USA
33. Filip PA, Balaban TS, Balaban AT (1987) A new approach for devising local graph invariants: derived topological indices with low degeneracy and good correlation ability. J Math Chem 1:61–83
34. Stewart JJP (1990) MOPAC Version 6.00, QCPE #455, Frank J Seiler Research Laboratory, US Air Force Academy, CO
35. Frisch MJ et al (1998) Gaussian 98 (Revision A.11.2). Gaussian, Inc., Pittsburgh
36. Auer CM, Nabholz JV, Baetcke KP (1990) Mode of action and the assessment of chemical hazards in the presence of limited data: use of structure-activity relationships (SAR) under TSCA, section 5. Environ Health Perspect 87:183–197
37. Yap CW (2011) PaDEL-descriptor: an open source software to calculate molecular descriptors and fingerprints. J Comput Chem 32:1466–1474
38. Todeschini R, Consonni V, Mauri A, Pavan M. (2006) DRAGON – Software for the calculation of molecular descriptors, version 5.4, Talete srl. Milan.
39. Johnson M, Basak SC, Maggiora G (1988) A characterization of molecular similarity methods for property prediction. Math Comput Mod 11:630–634
40. Tibshirani R (1996) Regression shrinkage and selection via the lasso. J R Stat Soc Ser B 58:267–288
41. Zou H (2006) The adaptive lasso and its oracle properties. J Am Stat Assoc 101:1418–1429
42. Cook RD, Li B, Chiaromonte F (2010) Envelope models for parsimonious and efficient multivariate linear regression. Stat Sin 20:927–1010
43. Hawkins DM, Basak SC, Mills D (2003) Assessing model fit by cross-validation. J Chem Inf Comput Sci 3:579–586
44. Hawkins DM, Basak SC, Mills D (2004) QSARs for chemical mutagens from structure: ridge regression fitting and diagnostics. Environ Toxicol Pharmacol 16:37–44
45. Basak SC, Mills D, Hawkins DM, Kraker JJ (2007) Proper statistical modeling and validation in QSAR: a case study in the prediction of rat fat-air partitioning. In: Simos TE, Maroulis G (eds) Computation in modern science and engineering, proceedings of the International Conference on Computational Methods in Science and Engineering 2007 (ICCMSE 2007). American Institute of Physics, Melville, pp 548–551
46. Basak SC, Majumdar S (2016) Current landscape of hierarchical QSAR modeling and its applications: Some comments on the importance of mathematical descriptors as well as rigorous statistical methods of model building and validation. In: Basak SC, Restrepo G, Villaveces JL (ed) Advances in mathematical chemistry and applications, vol 1. Bentham eBooks, Elsevier & Bentham Science Publishers, Sharjah, U. A. E, pp 251–281
47. Basak SC, Majumdar S (2015) Hierarchical quantitative structure-activity relationships (HiQSARs) for the prediction of physicochemical and toxicological properties of chemicals using computed molecular descriptors, Mol2Net Conference. http://sciforum.net/email/validate/49668c1bf65ab8520f721a84f7d84e05
48. Majumdar S, Basak SC, Grunwald GD (2013) Adapting interrelated two-way clustering method for quantitative structure-activity relationship (QSAR) modeling of mutagenicity/non-mutagenicity of a diverse set of chemicals. Curr Comput Aided Drug Des 9:463–471
49. Basak SC, Majumdar S (2015) Prediction of mutagenicity of chemicals from their calculated molecular descriptors: a case study with structurally homogeneous versus diverse datasets. Curr Comput Aided Drug Des 11:117–123
50. Basak SC, Majumdar S (2015) The importance of rigorous statistical practice in the current landscape of QSAR modelling (editorial). Curr Comput Aided Drug Des 11:2–4
51. Kraker JJ, Hawkins DM, Basak SC, Natarajan R, Mills D (2007) Quantitative structure-activity relationship (QSAR) modeling of juvenile hormone activity: comparison of validation procedures. Chemometr Intell Lab Syst 87:33–42
52. Hawkins DM, Kraker JJ, Basak SC, Mills D (2008) QSPR checking and validation: a case study with hydroxy radical reaction rate constant. SAR QSAR Environ Res 19:525–539
53. SAS Institute, Inc (1988) In SAS/STAT user guide, release 6.03 edition. Cary
54. Hoskuldsson A (1995) A combined theory for PCA and PLS. J Chemom 9:91–123

55. Hawkins DM, Basak SC, Shi X (2001) QSAR with few compounds and many features. J Chem Inf Comput Sci 41:663–670
56. Tang C, Zhang L, Zhang A, Ramanathan M (2001) Interrelated two-way clustering: an unsupervised approach for gene expression data analysis. In: Bilof R, Palagi L (eds) Proceedings of BIBE 2001: 2nd IEEE international symposium on bioinformatics and bioengineering, Bethesda, Maryland, November 4–5, 2001. IEEE Computer Society, Los Alamitos, pp 41–48
57. Basak SC, Magnuson VR, Niemi GJ, Regal RR, Veith GD (1987) Topological indices: their nature, mutual relatedness, and applications. Math Mod 8:300–305
58. Basak SC, Grunwald GD, Majumdar S (2015) Intrinsic dimensionality of chemical space: characterization and applications, Mol2Net conference. http://sciforum.net/email/validate/49668c1bf65ab8520f721a84f7d84e05
59. Basak SC (1999) Information theoretic indices of neighborhood complexity and their applications. In: Devillers J, Balaban AT (eds) Topological indices and related descriptors in QSAR and QSPR. Gordon and Breach Science Publishers, Amsterdam, pp 563–593
60. Randic M (1975) Characterization of molecular branching. J Am Chem Soc 97:6609–6615
61. Bonchev D, Trinajstić N (1977) Information theory, distance matrix and molecular branching. J Chem Phys 67:4517–4533
62. Hoffmann R, Minkin VI, Carpenter BK (1997) Ockham's razor and chemistry. HYLE Int J Philos Chem 3:3–28
63. Katritzky AR, Putrukhin R, Tathan S, Basak SC, Benfenati E, Karelson M, Maran U (2001) Interpretation of quantitative structure-property and -activity relationships. J Chem Inf Comput Sci 41:679–685
64. Katritzky AR, Putrukhin R, Tathan S, Basak SC, Benfenati E, Karelson M, Maran U (2001) Interpretation of quantitative structure-property and -activity relationships. J Chem Inf Comput Sci 41:679–685
65. So SS, Karplus M (1997) Three-dimensional quantitative structure-activity relationships from molecular similarity matrices and genetic neural networks. 2. Applications. J Med Chem 40:4360–4371
66. Basak SC, Mills D, Mumtaz MM, Balasubramanian K (2003) Use of topological indices in predicting aryl hydrocarbon (Ah) receptor binding potency of dibenzofurans: a hierarchical QSAR approach. Ind J Chem 42A:1385–1391
67. Basak SC, Majumdar S (2015) Current landscape of hierarchical QSAR modeling and its applications: some comments on the importance of mathematical descriptors as well as rigorous statistical methods of model building and validation. In: Basak SC, Restrepo G, Villaveces JL (eds) Advances in mathematical chemistry and applications, vol 1. Bentham eBooks, Bentham Science Publishers, pp 251–281
68. Ben-Dor A, Friedman N, Yakhini Z (2001) Class discovery in gene expression data. In: Proceedings of the fifth annual international conference on computational molecular biology (RECOMB 2001), New York
69. Gute BD, Basak SC (1997) Predicting acute toxicity of benzene derivatives using theoretical molecular descriptors: a hierarchical QSAR approach. SAR QSAR Environ Res 7:117–131
70. Gute BD, Grunwald GD, Basak SC (1999) Prediction of the dermal penetration of polycyclic aromatic hydrocarbons (PAHs): a hierarchical QSAR approach. SAR QSAR Environ Res 10:1–15
71. Basak SC, Mills DR, Balaban AT, Gute BD (2001) Prediction of mutagenicity of aromatic and heteroaromatic amines from structure: a hierarchical QSAR approach. J Chem Inf Comput Sci 41:671–678
72. Popper K (2005) The logic of scientific discovery. Taylor & Francis e-Library, London and New York
73. Basak SC, Majumdar S (2015) Two QSAR paradigms- congenericity principle versus diversity begets diversity principle- analyzed using computed mathematical chemodescriptors of homogeneous and diverse sets of chemical mutagens. Mol2Net Conference. http://sciforum.net/email/validate/49668c1bf65ab8520f721a84f7d84e05
74. Sahigara F, Mansouri K, Ballabio D, Mauri A, Consonni V, Todeschini R (2012) Comparison of different approaches to define the applicability domain of QSAR models. Molecules 17:4791–4810
75. Jaworska J, Nikolova-Jeliazkova N, Aldenberg T (2005) QSAR applicability domain estimation by projection of the training set descriptor space: a review. Altern Lab Anim 33:445–459
76. Preparata FP, Shamos MI (1991) Convex hulls: basic algorithms. In: Preparata FP, Shamos MI (eds) Computational geometry: an introduction. Springer, New York, pp 95–148
77. Worth AP, Bassan A, Gallegos A, Netzeva TI, Patlewicz G, Pavan M, Tsakovska I, Vracko M (2005) The characterisation of (quantitative) structure-activity relationships: preliminary guidance, ECB Report EUR 21866 EN. European Commission, Joint Research Centre, Ispra, p 95
78. Pharmaceutical Research and Manufacturers of America (2014) Biopharmaceutical research industry profile. Available from: http://www.phrma.org/sites/default/files/pdf/2014_PhRMA_PROFILE.pdf. Accessed on 11 Dec 2015
79. Santos-Filho OA, Hopfinger AJ, Cherkasov A, de Alencastro RB (2009) The receptor-dependent QSAR paradigm: an overview of the current state of the art. Med Chem (Shariqah) 5:359–366
80. Basak SC, Bhattacharjee AK, Vracko M (2015) Big data and new drug discovery: tackling "Big Data" for virtual screening of large compound databases. Curr Comput Aided Drug Des 11:197–201
81. Crawford MA (1963) The effects of fluoroacetate, malonate and acid-base balance on the renal disposal of citrate. Biochem J 8:115–120
82. Quastel JH, Wooldridge WR (1928) Some properties of the dehydrogenating enzymes of bacteria. Biochem J 22:689–702

83. Basak SC, Grunwald GD (1995) Predicting mutagenicity of chemicals using topological and quantum chemical parameters: a similarity based study. Chemosphere 31:2529–2546
84. Reuschenbach P, Silvani M, Dammann M, Warnecke D, Knacker T (2008) ECOSAR model performance with a large test set of industrial chemicals. Chemosphere 71:1986–1995
85. Ankley GT, Bennett RS, Erickson RJ, Hoff DJ, Hornung W, Johnson RD, Mount DR, Nichols JW, Russom CL, Schmeider PK, Serrano JA, Tietge J, Villeneuve DL (2010) Adverse outcome pathways: a conceptual framework to support ecotoxicology research and risk assessment. Environ Toxicol Chem 29:730–741
86. Ankley GT, Villeneuve DL (2006) The fathead minnow in aquatic toxicology: past, present and future. Aquat Toxicol 78:91–102
87. Basak SC, Grunwald GD, Host GE, Niemi GJ, Bradbury SP (1998) A comparative study of molecular similarity, statistical and neural network methods for predicting toxic modes of action of chemicals. Environ Toxicol Chem 17:1056–1064
88. Russom CL, Bradbury SP, Broderius SJ, Hammermeister DE, Drummond RA (1997) Predicting modes of toxic action from chemical structure: acute toxicity in the fathead minnow (pimephales promelas). Environ Toxicol Chem 16:948–967
89. Gute BD, Grunwald GD, Mills D, Basak SC (2001) Molecular similarity based estimation of properties: a comparison of structure spaces and property spaces. SAR QSAR Environ Res 11:363–382
90. Gute BD, Basak SC, Mills D, Hawkins DM (2002) Tailored similarity spaces for the prediction of physicochemical properties. Internet Electron J Mol Des 1:374–387. http://www.biochempress.com/
91. Basak SC, Gute BD, Mills D, Hawkins DM (2003) Quantitative molecular similarity methods in the property/toxicity estimation of chemicals: a comparison of arbitrary versus tailored similarity spaces. J Mol Struct THEOCHEM 622:127–145
92. Hamori E, Ruskin J (1983) H Curves, a novel method of representation of nucleotide series especially suited for long DNA sequences. J Biol Chem 258:1318–1327
93. Gates MA (1986) A simple way to look a DNA. J Theor Biol 119:319–328
94. Nandy A (1996) Graphical analysis of DNA sequence structure: III. Indications of evolutionary distinctions and characteristics of introns and exons. Curr Sci 70:661–668
95. Leong PM, Morgenthaler S (1995) Random walk and gap plots of DNA sequences. Comput Appl Biosci 11:503–507
96. Randić M, Zupan J, Balaban AT, Vikic-Topic D, Plavsic D (2011) Graphical representation of proteins. Chem Rev 111:790–862
97. Indo-US Workshop on Mathematical Chemistry. http://www.nrri.umn.edu/indousworkshop
98. Raychaudhury C, Nandy A (1998) Indexation schemes and similarity measures for macromolecular sequences. Paper presented at the Indo-US Workshop on Mathematical Chemistry, Shantiniketan. 9–13 January 1998
99. Randić M, Vracko M, Nandy A, Basak SC (2000) On 3–D representation of DNA primary sequences. J Chem Inf Comput Sci 40:1235–1244
100. Guo X, Randić M, Basak SC (2001) A novel 2-D graphical representation of DNA sequences of low degeneracy. Chem Phys Lett 350:106–112
101. Nandy A, Sarkar T, Basak SC, Nandy P, Das S (2014) Characteristics of influenza HA-NA interdependence determined through a graphical technique. Curr Comput Aided Drug Des 10:285–302
102. Nandy A, Basak SC (2015) Prognosis of possible reassortments in recent H5N2 epidemic influenza in USA: implications for computer-assisted surveillance as well as drug/vaccine design. Curr Comput Aided Drug Des 11:110–116
103. Steiner S, Witzmann FA (2000) Proteomics: applications and opportunities in preclinical drug development. Electrophoresis 21:2099–2104
104. Witzmann FA, Monteiro-Riviere NA (2006) Multiwalled carbon nanotube exposure alters protein expression in human keratinocytes. Nanomedicine Nanotechnol Biol Med 2:158–168
105. Basak SC, Gute BD, Monteiro-Riviere N, Witzmann FA (2010) Characterization of toxicoproteomics maps for chemical mixtures using information theoretic approach. In: Mumtaz M (ed) Principles and practice of mixtures toxicology. Wiley-VCH Verlag GmbH & Co. KGaA, Weinheim, pp 215–232
106. Vracko M, Basak SC, Geiss K, Witzmann FA (2006) Proteomics maps-toxicity relationship of halocarbons studied with similarity index and genetic algorithm. J Chem Inf Model 46:130–136
107. Randic M, Witzmann FA, Vracko M, Basak SC (2001) On characterization of proteomics maps and chemically induced changes in proteomes using matrix invariants: application to peroxisome proliferators. Med Chem Res 10:456–479
108. Basak SC, Gute BD, Witzmann FA (2006) Information-theoretic biodescriptors for proteomics maps: development and applications in predictive toxicology. Conf Proc WSEAS Trans Inf Sci Appl 7:996–1001
109. Arcos JC (1987) Structure–activity relationships: criteria for predicting the carcinogenic activity of chemical compounds. Environ Sci Technol 21:743–745
110. Hawkins DM, Basak SC, Kraker JJ, Geiss KT, Witzmann FA (2006) Combining chemodescriptors and biodescriptors in quantitative structure-

activity relationship modeling. J Chem Inf Model 46:9–16
111. Basak SC, Gute BD, Balaban AT (2004) Interrelationship of major topological indices evidenced by clustering. Croat Chem Acta 77:331–344
112. Johnson M, Maggiora GM (1990) Concepts and applications of molecular similarity. Wiley, New York
113. Basak SC (2016) Mathematical structural descriptors of molecules and biomolecules: background and applications. In: Basak SC, Restrepo G, Villaveces JL (ed) Advances in mathematical chemistry and applications, vol 1. Bentham eBooks, Elsevier & Bentham Science Publishers, Sharjah, U. A. E. pp 3–23
114. Zanni R, Galvez-Llompart M, Garcıa-Domenech R, Galvez J (2015) Latest advances in molecular topology applications for drug discovery. Expert Opin Drug Discov 10:1–13

Epigenetics Moving Towards Systems Biology

11

Arif Malik, Misbah Sultana, Aamer Qazi, Mahmood Husain Qazi, Mohammad Sarwar Jamal, and Mahmood Rasool

11.1 Introduction

The finding of DNA (Deoxyribonucleic acid) unfolded new era in the area of biotechnology and genomics. At present, genetics can precisely distinguish and influence the specific gene position inside genome which induces genetic disease, thus giving doorstep for possible cure of various diseases. Still, the basic function and structure of deoxyribonucleic acid is unable to explain the whole mechanisms of regulating gene and the development of disease. Nowadays, epigenetic is acquiring key stage to pursuit more beneficial understanding of genome and finally gene expression [1]. Epigenetic, an emerging area of biology, was initially specified in 1942 by Conrad Waddington, such phenomenon in which genes give rise to phenotype. Later on, in 1987, another scientist Robin Holliday added the DNA methylation patterns in the definition which affect the activity of gene [2]. At present, epigenetic is the field of changes in gene regulation which are not due to alterations in DNA sequence; genome can induce functionally applicable alterations which do not alter sequence of nucleotide. For many years, epigenetic has been assumed as a biological function [3]. On developmental stage, zygote begins in totipotent of which divided cells increasingly separate into myriad type of cells. This immensely give every cell a different type of phenotype in an individual, but all carry same genome e.g. the cell of eye is not like skin or neural cell. Genome, a complete set of genes or inherited material, contains genes and sequences of non-coding DNA. Epigenome had both histone-chromatin family (histones, DNA and DNA binding proteins) and patterns of DNA methylation. In 2008, epigenetic was demonstrated as 'stably inheritable phenotype' ensuing from chromosomal changes without modifications in Deoxyribonucleic Acid sequence [4].

The fundamental mechanisms of epigenetic modifications are complex and do methylation of DNA, histone modification and regulation of gene through non-coding RNAs [5, 6]. Further, epigenetic changes are transient and potentially reversible. These mechanisms can be affected by various environmental factors [7]. In the end, epigenetic modifications regulate expression of

A. Malik • M. Sultana
Institute of Molecular Biology and Biotechnology (IMBB), The University of Lahore, Lahore, Pakistan

A. Qazi • M.H. Qazi
Center for Research in Molecular Medicine (CRiMM), The University of Lahore, Lahore, Pakistan

M.S. Jamal
King Fahd Medical Research Center (KFMRC), King Abdulaziz University, Jeddah, Saudi Arabia

M. Rasool (✉)
Center of Excellence in Genomic Medicine Research (CEGMR), King Abdulaziz University, Jeddah, Saudi Arabia
e-mail: mahmoodrasool@yahoo.com

S. Singh (ed.), *Systems Biology Application in Synthetic Biology*,
DOI 10.1007/978-81-322-2809-7_11

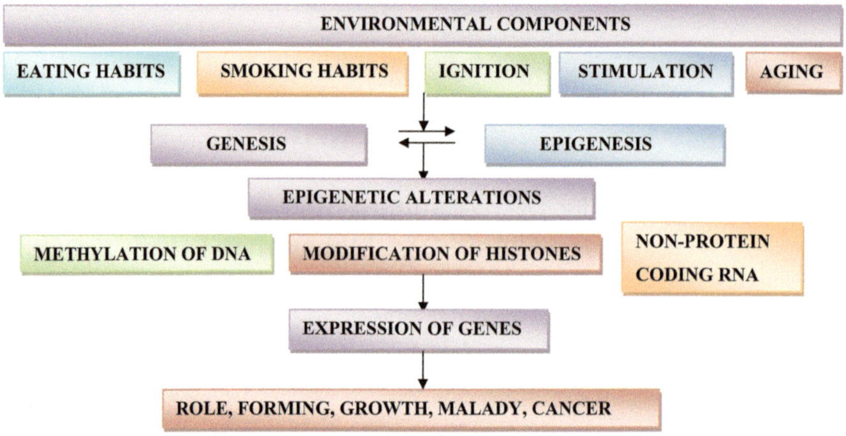

Fig. 11.1 Environmental components involved in epigenetic. Various environmental components like habit of smoking, eating habits, stimulation, ignition and aging might strike regulation of gene, that cause epigenetic alterations in genome. Mechanisms of epigenetic modifications are methylation of DNA, histone modification and regulation of gene through non-coding RNAs

gene and also affect many functions of gene (Fig. 11.1).

11.2 Mechanisms of Epigenetic

11.2.1 DNA Methylation

DNA methylation, named as "fifth base" of DNA, was acknowledged in 1948 [8]. DNA methylation gives short and semi-permanent consequences with expression of gene [9]. DNA methylation can specifically provoke epigenetic silencing of sequences like pluripotent-associated genes, transposons and impaired genes [10]. DNA methylation is one of the entire functions of various cellular processes, which includes development of embryo, genome forming, preserving chromosome consistency and inactivation of X-chromosome [11–13]. Scientists have achieved the insight of DNA methylation by how it occurs and target the sequence. The perturbation in epigenetics may cause complications like cancer or developmental problems [14]. Researchers have inter-related methylation of DNA and cancer [15]. Firstly, Feinburg and Vogelstein described methylation of DNA in human colon cancer and made comparison to normal cells [16]. Many preliminary analyses enhanced methylation of DNA importance in cells of cancer and predicted its function in other diseases and disorders.

11.2.2 DNA Methylation on Molecular Basis

DNA methylation, a process in which methyl group adds to 5 carbon of cytosine which yields 5-mC. DNA methylation takes place in circumstance of cytosine which introduces guanine [17]. Guanines are extremely interpreted in genome; however 70 % of them are methylated and other are unmethylated, often present in "guanine islands". Guanine islands are part of genome which constitutes 200 bp in length [18]. Mostly an increase ratio of guanine characterizes 60 % of human promoters as guanine is fertilized in 5′ promoter area of genes [19]. Even so, guanine concentration does not regulate gene expression. Rather, transcriptional regulation depends much upon DNA methylation position. Generally, CpG (guanine) islands which are promoter-associated at the stage of transcriptionally active genes remain unmethylated [18]. For the first time, it was demonstrated that silencing of gene takes place in diploid somatic cells through methylation (apart from inactivation of X-chromosome) comprised of malignant tumor gene suppressor

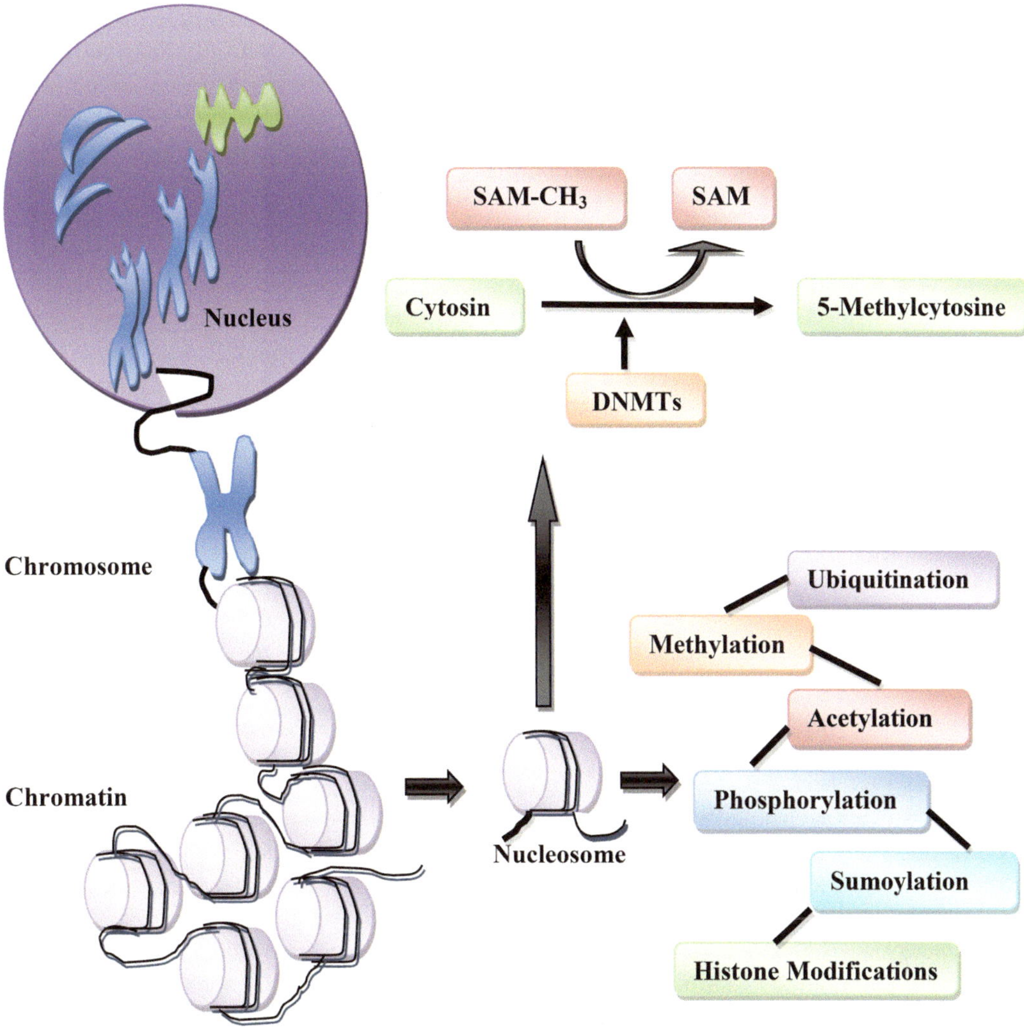

Fig. 11.2 Schematic of epigenetic alterations. Strands of DNA are enfolded across histone octamers, thus nucleosome forms which organize within chromatin. Chromatin is the building blocks of chromosome. DNMTs from methyl donor group transfers SAM to 5-methylcytosine. Reversible histone alterations take place through ubiquitination, acetylation, phosphorylation, methylation and sumoylation

[14]. Subsequently, various tumor gene suppressor constituted to silencing through mechanisms of epigenetic [18].

The reaction of methylation which impart 5′ cytosine moiety is catalyzed through DNA methyltransferases (DNMTs) enzymes. Such enzymes take methyl radical from S-adenosylmethionine (SAM) donor and transfer it to 5′ cytosine. (Fig. 11.2). Family of DNMT constitutes on five members, which includes DNA methyltransferase 1, DNA methyltransferase 2, DNA methyltransferase 3a, 3b and 3 L [20]. DNA methyltransferase 1, 3a and 3b act on cytosine base to give global methylation or methylome. These are further separated as de novo DNA methyltransferease 3a and 3b or DNA methyltransferase1 maintenance enzymes. DNA methyltransferase 2 and 3 L could not act as CMT (cytosine methyltransferase) [18]. DNA methyltransferase 3 L, having similarity with DNMTs3a induces de novo DNA methylation action by enhancing the binding affinity with S-adenosylmethionine, along with mediation of transcriptional repressor gene by inscribing histone deacetylase 1 [21–23]. DNA methyltransferase does not own N-terminal regulatory domain just like other DNA methyltransferse enzymes. It is believed that DNA methyltransferasae may be needed for DNA damaging and repairing response [24].

DNA methlytransferase 1 impart methylation of template parental DNA strand to daughter DNA strand when replication of DNA occurs. This assures same methylome in the leading cells. Such activity is needed for proper functioning of cell and methylation maintenance during somatic cell division. DNA methyltransferase 3a and 3b accomplished de novo DNA methylation throughout embryogenesis and development of germ cell [25]. It was observed that 5-hmC (5-hydroxymethylcytosine) formed by the oxidation of 5-methyl cytosine (5-mC) through TET (ten-eleven translocations) proteins. 5-Hydroxymethylcytosine is structurally same like 5-methylcytosine, and at the beginning it was observed in embryonic stem cells and cerebellar neurons [26–28]. Many other mechanisms have been discovered which substitute 5-methylcytosine onto unmethylated cytosine and make 5-hydroxymethylcytosine by ten-eleven translocation enzymes, at last DNA gylcosylase enzyme family repairs the base excision [29]. 5-Methylcytosine can be changed through ten-eleven translocation proteins into 5-formylcytosine and 5-carboxylcytosine during demethylation of DNA [30]. The distinct function of DNA methyltransferase have been focused for further research findings and among them epigenetic has been discovered [31]. In fact, in vitro condition DNA methyltransferase 3a and 3b can act as dehydroxymethylases and DNA methyltransferases [32].

11.2.3 Histone Posttranslational Modifications

Basically, the amino end tails of core histones, i.e. H2A, H2B, H3 and H4, are reactive and sensory to various modifications which includes methylation, ubiquitination, acetylation, sumoylation and phosphorylation [33, 34]. In spherical cores, histones are strongly packed to N-terminal amorphous tails which project outwards. Histone-modifying enzymes target by these tails. Finally, at full extension, N-terminal histone tails extends substantially outside the super helical turns of DNA [35]. The histone tails are very rich within lysine residues which are extremely charged positively at physiological pH [36]. The positively charged lysine bind to negatively charged DNA tightly, as a result nucleosomes get condense and structure of chromatin forms which is transcription factor cannot access. Histone modifications, type of posttranslational modifications, are necessary to control structure and function of chromatin that affects DNA-linked processes like transcription and organization of chromosomes [37]. The most dominant posttranslational modifications along heterochromatin euchromatin are methylation and acylation of lysine residues present at tails of histone [38]. Histone acetyltransferases (HATs) catalysis histone lysine acetylation, and thus positively charged histone tails are neutralized by acetyl group while histones affinity decreases for negatively charged DNA. The DNA and histones association loses, hence facilitates transcription factors to access promoter regions and therefore transcriptional activity increases [39–42].

Among epigenetic modifications, for the first time histone acetylation was correlated to regulation of transcriptions [43–45]. Activation of gene against transcriptional repression is achieved by changes in between histone acetyltransferase (HAT) and activities of histone deacetylase (HDAC), respectively [46]. The function of these enzymes is in mutliprotein complexes which modulate chromatin in extremely particular ways. Acetyl group transfers from acetyl CoA to amino radical of lysine residues through histone acetyltransferases with coenzyme-A as the final product. Researchers suggest that protein-protein interactions get site from lysine acetylation, such as acetyl lysine-binding bromodomain and results in soft euchromatin configuration [47–50]. Histone acetyltransferase had three main classes i.e. GNATs (Gcn5-related N-acetyltransferase), MYST and p300/CBP [51, 52]. Bromodomain characterized Gcn5-related N-acetyltransferase

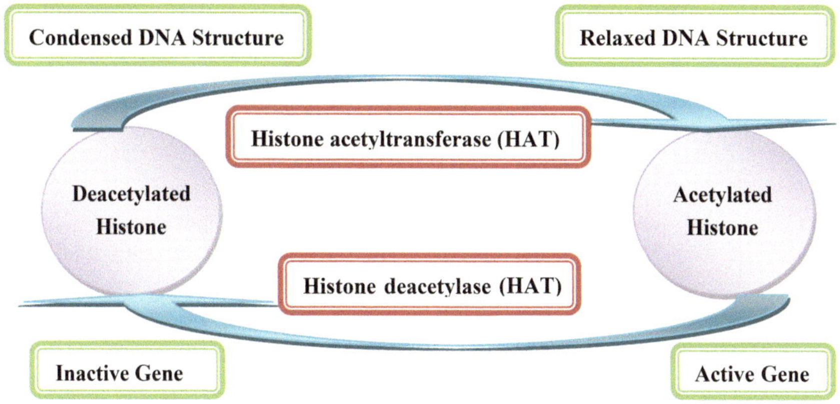

Fig. 11.3 Schematic representation of reversible alterations in chromatin. Genes activated when DNA structure is open while genes inactivated when DNA structure in condensed

through which lysine residues acetylates on H2B, H3 and H4 [53]. The four members' family MYST acetylates the lysine residues with H2A, H3 and H4 while p300/CBP acetylate lysine with all four histones H2A, H2B, H3 and H4.

Histone deacetylase catalysis the reverse reaction by raising the positive charge present on histone tails, thus transcriptional potential from under-lysine gene get hindered through close binding to negatively charged DNA. In fact, in biological systems it is substantially known that loci repressed transcriptionally area linked to deacetylated histones [54–56]. Histone deacetylase are of many kinds which on the basis of sequence and function constitute four groups similar to yeast protein. The group 1 and group 2 primarily comprise of members which are classically zinc-dependent. Group 1 contains histone deacetylase 1, 2, 3 and 8. Histone deacetylase 1, 2 and 8 are placed primarily within nucleus, as histone deacetylase 3 is established in nucleus, cytoplasm and also associated with membrane. Group 2 includes histone deacetylase 4, 5, 6, 7, 9 and 10 that in response to particular signal, transport in and out of nucleus [57, 58]. These two group deacetylate lysine which plays an important role in inactivation of transcription [59].

Methylation of histones has been reported as the fundamental, differentiating, epigenetic figure associated with gene activity [60, 61], while histone hyperacetylation is correlated positively to actively transcribed genes [62]. Histone methylation is correlated cellularly with DNA replication and repairing. Within these, repression and transcriptional activation area mostly analyzed [59]. Histones are only methylated in lysine/arginine residues from histone tails H3 and H4 [63]. However, methylated histone is mostly found in lysine residue (Fig. 11.3). Chromatin figure changes by methylation not only by changing the charge on lysine residue but also by elevating and limiting the docking of chromatin linked proteins and transcriptional factors. Generally, methylated histone is enriched with activated regions of gene, especially at K4, K36 or K79 [64–66]. On the other side, methylation enriched at lysine residues K9, K20 or K27 has been concerned in inactivation and silencing of gene [34]. Amino group is present in both arginine and lysine residues which confer main hydrophobic features. Lysine could be mono, di or trimethylated but as far as arginine is concerned it might be mono or dimethylated. Many cofactors and substrates with various enzymes are needed for methyl group to attach with residue. Protein arginine methyltransferase is required for arginine methylation while histone methylation is involved in lysine methylation.

Histone methyltransferase enzymes are enzymes in which SAM transfers methyl group onto lysine and/arginine. Various covalent modifications found in histone tails could reverse enzymatically e.g. deacetylase and phosphate can reverse acetylation and phosphorylation. This enables the cell to react quickly to modifications inside cellular surroundings through rapidly modifying the regulatory gene machinery. In 1960s, scientists discovered histone lysine methylation static [67–69]. Later on in 2004, histone demethylsae lysine-specific demethylase1 (LSD1) was discovered which demonstrated histone lysine methylation to be dynamic [70]. Since then, linker and core histones have been cataloged as sites of methylation and identified such enzymes which catalysis gain or removal of methyl group [71]. The position of histone lysine methylation is regulated by KMTs (lysine methyltransferases) and KDMs (demethylases). The substrates of non-histone are targeted by lysine methyltransferases [72, 73].

The two main types of lysine demethylases which utilize oxidative mechanisms are 2-oxoglutarate-(2OG) dependent JmjC and the flavin-dependent (LSDs) subfamily [74, 75]. Lysine monomethylated and dimethylated residues can be demethylated by the flavin-dependent demethylase (LSDs). The arginine and lysine abundance on tails of histone combined to the various potential offers tremendous regulatory potential. Discovery of histone demethylases had notable effects on epigenetic. Surely, it has been proved that methylation of histone is reversible, still scientists are working to search other demethylases [28].

Another type of posttranslational alteration is histone phosphorylations that is involved in regulation of transcription and also do compression of chromatin [76]. Each histone tail has its own accepter site that get phosphorylated through protein kinases and phosphatases dephosphorylated. Expression of gene is through phosphorylated histones, especially regulation of growing genes. Further, histone H3S10 phosphorylation has been linked with acetylation of histone H3, strongly entailing such alterations in activation of transcription [77]. Histone phosphorylation also functions in compaction of chromatin. In the beginning found to be linked to compaction of chromosome throughout meiosis and mitosis, phosphorylation of histone H3 is also needed for regulating and relaxing gene expression in chromatin [78–80].

Many other histone tails posttranslational modifications includes sumoylation, ubiquitination and propionylation are also acknowledged and further crosstalk is going on about different histone modifications which change according to the environment changes. The active position of epigenetic modifications can influence chromatin which favors on (euchromatin) and off (heterchromatin) state [60].

11.2.4 Chromatin

Chromatin relates with the DNA complex and also with histone proteins which form genome. Genome is about 2 m long. Nucleosomes, the main building block of chromatin, are formed when DNA transfers over histone proteins. It is the first compaction stage in which DNA fits within the nucleus in organized way. Nucleosomes comprise of four proteins known as histones. Histones are known as H2A, H2B, H3 and H4. Another type of histone is H1 also called as linker histone. H1 (linker histone) binds with DNA within nucleosomes, and thus stabilizes and facilitates the nucleosomes to organize high order structure of chromatin [81, 82]. Due to this chromatin organization, DNA packaged tightly, also replicate properly and during cell division classified into daughter cells (Fig. 11.3).

Chromatin within non-dividing cell is further classified into heterochromatin and euchromatin that is transcriptionally inactive or active state of chromosome [33, 38] (Table 11.1). Euchromatin is the area in which DNA is approachable while in heterochromatin as DNA is tightly packed so is inaccessible for transcription factors [83]. Euchromatin had flexible genomic areas and genes are present in both active and inactive transcriptional state. Conversely, heterochromatin had genomic regions which comprise of insistent sequences and genes are linked to morphogenesis

Table 11.1 Epigenetic modifications influences chromatin status into two states: on (euchromatin)/off (heterochromatin). Methylation of DNA and modifications of epigenetic is exemplified in this table. Among the silencing effects of gene with modifications, H3K9me3 plays critical role in formation of heterochromatin. Still, it is not completely understood by which means these different epigenetic modifications are generated and asserted

Chromatin features		Heterochromatin	Euchromatin
	Structure	Condensed, closed, inaccessible	Less condensed, open, accessible
	Activity	DNA expression silenced	Active DNA expression
	DNA sequence	Repetitive elements	Gene rich
Epigenetic markers	DNA methylation	Hypermethylated	Hypomethylated
	Histone acetylation	Hypoacetylated at H3 and H4	Hyperacetylated at h3 and h4
	Histone methylation	H3K27me2,	H3K4me2,
		H3K27me3,	H3K4me3,
		H3K9me2,	H3K9me1
		H3K9me3	

[84]. Heterochromatin plays an important role in stability of chromosome and also prevents translocations and mutations [85].

At present, chromatin not only functions in package of DNA and regulation on inherited information but also activates the structure of chromatin and controls the function of genome to further determine the cellular behavior [86]. The distribution of epigenetic markers along with high-order functional areas is represented by chromatin territories (Table 11.1) [37]. Various epigenetic mechanisms regulate active composition of chromatin throughout the cell cycle. However, the high-order formation, regulation of chromatin and their effect on activity of genome is still elusive.

11.2.5 Non-Protein Coding RNAs

Non-protein coding RNAs are molecules of ribonucleic acid which are not interpreted into protein. Non-protein coding RNAs include ribosomal RNAs (rRNAs), short-interfering ribonucleic acids (siRNAs), transfer RNAs (tRNAs) and microRNAs (miRNAs). Regulation of gene expression is through microRNAs and short-interfering RNAs without changing the sequence of DNA. For example, at posttranscriptional level, micro RNAs which are 20–24 nucleotides small single-stranded moelcules regulate negatively the targeted genes expression [5, 6]. Micro RNAs can inhibit the expression of mRNA after binding to its target through various mechanisms. Although translational repression is one of the common mechanisms which occurs due to the binding of micro RNA to 3′ unstranslated region of mRNA. Guo et al. proposed that destablization of target mRNA enable the endogenous microRNAs to reduce protein level. Recently, it has been reported that microRNAs are found to be involved in various processes, during differentiation and developmental regulation of disease [87].

11.3 Role of Epigenetics

Scientists are actively participating to study the epigenetic modifications occurring throughout the initiation, growth and metastatic levels of cancer, in order to help the patient by developing improve diagnostic tools and therapeutic treatment. Epigenetic modifications also occur throughout fetal growth, cancer progression or within chronic diseases like diabetes mellitus, autoimmune, mental and cardiovascular in grownups [88]. Epigenetic mechanisms associated with the regulation of gene are discussed in the following section (Fig. 11.1).

11.3.1 Forming

Diploid beings inherit two gene copies, one from each parent. Researchers have proposed that inherited genes from each parent have been permanently differentiated and imprinted [89]. Thus, expression pattern which depends on inheritance of parental and maternal will demonstrate a mosaic pattern of parents. In mammals, imprinting of genome mediates that alleles expression through certain loci of gene is not equivalent rather is influenced through parent origin [90]. For instance, investigators discovered that H19 and IGF2R (Insulin-like growth factor-2 receptor) are merely activated if transmitted from mother, while expression of insulin-like growth factor-2 is just passed from father.

Methylation of DNA is one of the main underlying mechanisms of impressing. On this procedure, one gene imitate is marked on methylation of DNA which depends upon maternal source. During cell division, DNA methylation is asserted through 5-cytosine DNMT1 (DNA methytransferase-1) [91, 92]. DNA methyltransferase-1 expresses methylation inside the hemimethylated guanine (CpG) region and thus such methylated regions replicate to synthesize new strands of DNA. The best example of imprinting is insulin growth factor-2 which is regulated on fetal development [89]. For fetus somatic growth, insulin growth factor-2 is considered to be essential factor and any impairment could lead to damaging results. Thus, epigenetic platform through which insulin growth factor-2 (IGF2) gene expression is regulated is the main constituent of proper development.

11.3.2 Growth

Somatic epigenetic hereditary pattern such as methylation of DNA and remodeling of chromatin patterns is the very essential for the growth of multicellular eukaryotic organisms. Though sequence of gene is stable, yet differentiations of cells occur in many ways. They contain different functions and divergently react with the environment and also with intracellular signaling. Thus, epigenetic mechanisms play key role in performing different cellular functions and differentiation.

Recently, it has been described that regulation of gene expression by cell lineages is through epigenetic mechanism. For instance, epigenetic program regulates T-helper cell from immune system [93]. As T-cells (CD4+) become mature, it epigenetically activates interferon gamma (IFNγ) gene and silences interleukin-4 (IL-4) gene. This mechanism contributes to improper responses of T-cell, as actions of antigen and cytokine alter the epigenetic modification. Thus, different T-helper cells are formed to assert a polarized phenotype.

11.3.3 Environmental Components

Environmental factors can begin the alterations in DNA methylation as soon as the maternal stage. For instance, fetal DNA methylation is modified because of decrease level of dietary folate, or methionine in utero, and can persist substantially in adulthood [94]. Barker et al. reported that intrauterine exposures can induce fetus programming which lasts into adulthood and thus raise the risk of adult problems like diabetes mellitus type-2 and cardiovascular disease [95]. Thus, nutrition of intrauterine significantly affects the fetal epigenetic programming. For instance, the important methyl donor of S-adenosylmethyltransferase (SAM) is methyltetrahydrofolate that is used through enzyme, DNA methyltransferase, to further methylate guanine (CpG) residues [96]. During pregnancy, deficiency of folate in mother leads to poor level of S-adenosylmethyltransferase (SAM) [91]. Therefore, deficiency of folate in maternal can cause DNA hypomethylation that leads to excessive gene expression and genetic imbalancing in fetus [96]. Additionally, during life many environmental and dietary factors determine the epigenetic alterations.

11.3.4 Ignition

Ignition is a biological reaction for noxious stimuli like irritants and pathogens. Various studies proposed that epigenetic modifications are due to inflammation which includes methylation of DNA, histone modification and targeting through miRNAs [7]. Ito suggested that the action of nuclear component kappa-light-chain enhanced from the activation of B cells (NF-kB) is promoted by incitive signals, thus promotes the expression of gene and modifies histone methylation [97].

11.3.5 Cancer

During cancer, the well-known epigenetic alteration observed is DNA methylation. These epigenetic modifications are assorted as the main components of carcinogenesis. Mostly hypomethylation takes place in tumor that raises transcriptional activity. This might take place in unstable sequence and is associated with raised frequency of tumor. It has been considered as the earlier epigenetic alteration intending to change cells from normal to pre-malignant stage [89]. A few researches observed that hyper-methylation from neoplasm suppressor gene is associated with carcinogenesis [98]. Hyper-methylation for neoplasm-suppressor genes causes repression of genes and subsequently leads to progression of tumor [99]. It has been reported that epigenetic modification may originate oncogenesis. Though researches are being made on epigenetics, various studies have highlighted the effects of epigenetic on health and also contributing in the development of regenerative treatment [100–102].

11.4 Conclusion and Future Perspectives

Epigenetics plays the key role in regulation of gene. Mechanisms relevant to epigenetic include methylation of DNA, modification of histone and non-protein coding RNAs. Although functions from these mechanisms are altered, still expression of gene is affected by them. Epigenetic alteration can lead to imprinting of gene and causes development of regulation among the eukaryotic organisms. Moreover, exogenic factors like smoking, inflammation, diet and stimuli can lead epigenetic changes regulated by expression of gene. Epigenetic modifications can lead to certain disease progression like cancer. Today, epigenomic is considered as the most exciting region in biomedicine. Epigenetic mechanism detected in health and disease not only provides understanding about the origins of human malady but also gives framework for developing new medical aids.

References

1. Seo JY, Park YJ, Yi YA, Hwang JY, Lee IB, Cho BH, Son HH, Seo DG (2015) Epigenetics: general characteristics and implications for oral health. Restor Dent Endod 40:14–22
2. Holiday R (1987) The inheritance of epigenetic defects. Science 238:163–170
3. Goldberg AD, Allis CD, Bernstein E (2007) Epigenetics: a landscape takes place. Cell 128:635–638
4. Berger SL, Kouzarides T, Shiekhattar R, Shilatifard A (2009) An operational definition of epigenetics. Genes Dev 23:781–783
5. Bayarsaihan D (2011) Epigenetic mechanism in inflammation. J Dent Res 90:9–17
6. Kaikkonen MU, Lam MT, Glass CK (2011) Non-coding RNAs as regulators of gene expression and epigenetics. Cardiovasc Res 90:430–440
7. Lod S, Johansson T, Abrahamsson KH, Larsson L (2014) The influence of epigenetics in relation to oral health. Int J Dent Hyg 12:48–54
8. Doerfler W, Toth M, Kochanek S, Achten S, Freisem-Rabien U, Behn-Krappa A (1990) Eukaryotic DNA methylation: facts and problems. FEBS Lett 268:329–333
9. Bonasio R, Tu S, Reinberg D (2010) Molecular signals of epigenetics states. Science 330:612–680
10. Li G, Reinberg D (2011) Chromatin higher-order structures and gene regulation. Curr Opin Genet Dev 21:175–186
11. Efstratiadis A (1994) Parental imprinting of autosomal mammalian genes. Curr Opin Genet Dev 4:265–280
12. Li E, Beard C, Jaenisch R (1993) Role for DNA methylation in genomic imprinting. Nature 366:362–365

13. Kundaje A, Meuleman W, Ernst J, Bilenky M, Yen A, Heravi-Moussavi A (2015) Integrative analysis of 111 reference human epigenomes. Nature 518:317–330
14. Esteller M (2008) Epigenetics in cancer. N Engl J Med 358:1148–1159
15. Gama-Sosa MA, Slagel VA, Trewyn RW, Oxahandler R, Kuo KC, Gehrke CW (1983) The 5-methylcytosine content of DNA from human tumors. Nucleic Acids Res 11:6883–6894
16. Feinberg AP, Vogelstein B (1983) Hypomethylation distinguishes genes of some human cancers from their normal counterparts. Nature 301:89–92
17. Portela A, Esteller M (2010) Epigenetic modifications and human disease. Nat Biotechnol 20:1057–1068
18. Kulis M, Esteller M (2010) DNA methylation and cancer. Adv Genet 70:27–56
19. Saxonov S, Berg P, Brutlag DL (2006) A genome-wide analysis of CpG dinucleotides in the human genome distinguishes two distinct classes of promoters. Proc Natl Acad Sci U S A 103:1412–1417
20. Jurkowska RZ, Jurkowski TP, Jeltsch A (2011) Structure and function of mammalian DNA methyltransferases. Chembiochem 12:206–222
21. Chedin R, Lieber MR, Hsieh CL (2002) The DNA methyltransferase-like protein DNMT3L stimulates de novo methylation by DNMT3a. Proc Natl Acad Sci U S A 99:16916–16921
22. Deplus R, Brenner C, Burgers WA, Putmans P, Kouzarides T, de Launoit Y (2002) DNMT3L is a transcriptional repressor that recruits histone deacetylase. Nucleic Acids Res 30:3831–3838
23. Aapola U, Liiv I, Peterson P (2002) Imprinting regular DNMT3L is a transcriptional repressor associated with histone deacetylase activity. Nucleic Acids Res 30:3602–3608
24. Subramaniam D, Thombre R, Dhar A, Anant S (2014) DNA methyltransferases: a novel target for prevention and therapy. Front Oncol 4:80
25. Law JA, Jacobsen SE (2010) Establishing maintaining and modifying DNA methylation patterns in plants and animals. Nat Rev Genet 11:204–220
26. Tahiliani M, Koh KP, Shen Y, Pastor WA, Bandukwala H, Brudno Y (2009) Conversion of 5-methylcytosine to 5-hydroxymethycytosine in mammalian DNA by MLL partnet TET1. Science 324:930–935
27. Kriaucionis S, Heintz N (2009) The nuclear DNA base 5-hydroxymethylcytosine is present in Purkinje neurons and the brain. Science 324:929–930
28. Kriaucionis S, Tahiliani M (2014) Expanding the epigenetic landscape: novel modifications of cytosine in genomic DNA. Cold Spring Harb Perspect Biol 6:a018630
29. Li CJ (2013) DNA demethylation pathways: recent insights. Genet Epigenet 5:43–49
30. Ito S, Shen L, Dai Q, Collins LB, Wu SC, Swenberg JA (2011) Tet proteins can convert 5-methylcytosine to 5-formylcytosine and 5-carboxylcytosine. Science 333:1300–1303
31. Murr R (2010) Interplay between different epigenetic modifications and mechanisms. Adv Genet 70:101–141
32. Chen CC, Wang KY, Shen CK (2012) The mammalian de novo DNA methyltransferases DNMT3A and DNMT3B are also DNA 5-hydroxymethylcytosine dehydroxymethylases. J Biol Chem 287: 33116–33121
33. Kouzarides T (2007) Chromatin modifications and their function. Cell 128:693–705
34. Ruthenburg AJ, Li H, Patel DJ, Allis CD (2007) Multivalent engagement of chromatin modifications by linked binding modules. Nat Rev Mol Cell Biol 8:983–994
35. Luger K, Mader AW, Richmond RK, Sargent DF, Richmond TJ (1997) Crystal structure of the nucleosome core particle at 2.8 A resolution. Nature 389:251–260
36. Urnov FD, Wolffe AP (2001) Above and within the genome epigenetics past and present. J Mammary Gland Biol Neoplasia 6:153–167
37. Koch CM, Andrews RM, Flicek P, Dillon SC, Karaoz U, Clelland GK (2007) The landscape of histone modifications across 1% of the human genome in five in five human cell lines. Genome Res 17:691–707
38. Jenuwein T, Allis CD (2001) Translating the histone code. Science 293:1074–1080
39. Vermaak D, Steinbach OC, Dimitrov S, Rupp RA, Wolffle AP (1998) The globular domain of histone H1 is sufficient to direct specific gene expression in early Xenopus embryos. Curr Biol 8:533–536
40. Rothbart SB, Strahl BD (2014) Interpreting the language of histone and DNA modifications. Biochim Biophys Acta 1839:627–643
41. Cheung WL, Briggs SD, Allis CD (2000) Acetylation and chromosomal functions. Curr Opin Cell Biol 12:326–333
42. Wolffe AP, Hayes JJ (1999) Chromatin disruption and modification. Nucleic Acids Res 27:711–720
43. Grunstein M (1997) Histone acetylation in chromat in structure and transcription. Nature 389:349–352
44. Brownell JE, Allis CD (1996) Special HATs for special occasions: linking histone acetylation to chromatin assembly and gene activation. Curr Opin Genet Dev 6:176–184
45. Wade PA, Wolffe AP (1997) Histone actyltransferases in control. Curr Biol 7:R82–R84
46. Yang XJ, Seto E (2008) The Rpd3/Hdal family of lysine deacetylases: from bacteria and yeast to mice and men. Nat Rev Mol Cell Biol 9:206–218
47. Yang XJ (2004) Lysine acetylation and the bromodomain: a new partnership for signaling. Bioessays 26:1076–1087

48. Mujtaba S, Zeng L, Zhou MM (2007) Structure and acetyl-lysine recognition of the bromodomain. Oncogene 26:5521–5527
49. Dhalluin C, Carlson JE, Zeng L, He C, Aggarwal AK, Zhou MM (1999) Structure and ligand of a histone acetyltransferase bromodomain. Nature 399:491–496
50. Dyson MH, Rose S, Mahadevan LC (2001) Acetylation-binding and function of bromodomain-containing proteins in chromatin. Front Biosci 6:D853–D865
51. Roth SY, Denu JM, Allis CD (2001) Histone acetyltransferases. Annu Rev Biochem 7:81–120
52. Marmorstein R, Roth SY (2001) Histone acetyltransferases: function, structure and catalysis. Curr Opin Genet Dev 11:155–161
53. Lee KK, Workman JL (2007) Histone acetyltransferase complexes: one size dosen't fit all. Nat Rev Mol Cell Biol 8:284–295
54. Jeppesen P, Turner BM (1993) The inactive X chromosome in female mammals is distinguished by a lack of histone H4 acetylation, a cytogenetic marker for gene expression. Cell 74:281–289
55. Braunstein M, Rose AB, Holmes SG, Allis CD, Broach JR (1993) Transcriptional silencing in yeast is associated with reduced nucleosome aacetylation. Genes Dev 7:592–604
56. Rundlett SE, Carmen AA, Suka N, Turner BM, Grunstein M (1998) Transcriptional repression by UME6 involves deacetylation of lysine 5 of histone H4 by RPD3. Nature 392:831–835
57. de Rujiter AJ, van Gennip AH, Caron HN, Kemp S, van Kuilenburg AB (2003) Histone deacetylases (HDACs): characterization of the classical HDAC family. Biochem J 370:737–749
58. Longworth MS, Laimins LA (2006) Histone deacetylase 3 localizes to the plasma membrane and is a substrate of Srcc. Oncogene 25:4495–4500
59. Zhang G, Pradhan S (2014) Mammalian epigenetic mechanisms. IUMBM Life 66:240–256
60. Berger SL (2007) The complex language of chromatin regulation during transcription. Nature 447:407–412
61. Guenther MG, Levine SS, Boyer LA, Jaenisch R, Young RA (2007) A chromatin landmark and transcription initiation at most promoters in human cells. Cell 130:77–88
62. Roh TY, Cuddapah S, Zhao K (2005) Active chromatin domains are defined by acetylation islands revealed by genome-wide mapping. Genes Dev 19:542–552
63. Wood A, Shilated A (2004) Posttranslational modifications of histones by methylation. In: Ronald CC, Joan Weliky C (eds). Adv protein chem 67:201–222
64. Heintzman ND, Stuart RK, Hon G, Fu Y, Ching CW, Hawkins RD (2007) Distinct and predictive chromatin signatures of transcriptional promoters and enhancers in the human genome. Nat Genet 39:311–318
65. Edmunds JW, Mahadevan LC, Clayton AL (2008) Dynamic histone H3 methylation during gene induction: HYPB/Setd2 mediates all H3K36 trimethylation. EMBO J 27:406–420
66. Steger DJ, Leftrova MI, Ying L, Stonestrom AJ, Schupp M, Zhuo D (2008) DOT1L/KMT4 recruitment and H3K79 methylation are ubiquitously coupled with gene transcription in mammalian cells. Mol Cell Biol 28:2825–2839
67. Pedersen MT, Helin K (2010) Histone demethylases in development and disease. Trends Cell Biol 20:662–671
68. Allfrey VG, Mirsky AE (1964) Structural modifications of histones and their possible role in the regulation of RNA synthesis. Science 144:559
69. Murray K (1964) The occurrence of epsilon-N-methyl lysine in histones. Biochemistry 3:10–15
70. Shi Y, Lan F, Matson C, Mulligan P, Whetstine JR, Cole PA (2004) Histone demethylation mediated by the nuclear amine oxiadase homolog LSD1. Cell 119:941–953
71. Greer EL, Shi Y (2012) Histone methylation: a dynamic mark in health, disease and inheritance. Nat Rev Genet 13:343–357
72. Moore KE, Gozani O (2014) An unexpected journey: lysine methylation across the proteome. Biochim Biophys Acta 1839:1395–1403
73. Clarke SG (2013) Protein methylation at the surface and buried deep: thinking outside the histone box. Trends Biochem Sci 38:243–252
74. Thinnes CC, England KS, Kawamura A, Chowdhury R, Schofield CJ, Hopkinson RJ (2014) Targeting histone lysine demethylases-progress, challenges and the future. Biochim Biophys Acta 1839:1416–1432
75. Shi YG, Tsukada Y (2013) The discovery of histone demethylases. Cold Spring Harb Perspect Biol 3:5
76. Rossetto D, Avvakumov N, Cote J (2012) Histone phosphorylation : a chromatin modification involved in diverse nuclear events. Epigenetics 7:1098–1108
77. Lo WS, Trievel RC, Rojas JR, Duggan L, Hsu JY, Allis CD (2000) Phosphorylation of serine 10 in histone H3 is functionally linked in vitro and in vivo to Gcn5-mediated acetylation at lysine 14. Mol Cell 5:917–926
78. Wei Y, Mizzen CA, Cook RG, Gorovsky MA, Allis CD (1998) Phoshorylation of histone H3 at serine 10 is correlated with chromosome condensation during mitosis and meiosis in Tctrehymena. Proc Natl Acad Sci U S A 5:7480–7484
79. Suave DM, Anderson HJ, Ray JM, James WM, Roberge M (1999) Phosphorylation-induced rearrangement of the histone H3 NH2-terminal domain during mitotic chromosome condensation. J Cell Biol 145:225–235
80. de la Barre AE, Gerson V, Gout S, Creaven M, Allis CD, Dimiitrov S (2000) Core histone N-terminal play an essential role in mitotic chromosome condensation. EMBO J 19:379–391

81. Shen X, Yu L, Weir JW, Gorovsky MA (1995) Linker histones are not essential and affect chromatin condensation in vivo. Cell 82:47–56
82. Dasso M, Dimitrov S, Wolffe AP (1994) Nuclear assembly is independent of linker histones. Proc Natl Acad Sci U S A 91:12477–12481
83. Talbert PB, Henikoff S (2006) Spreading of silent chromatin: inaction at a distance. Nat Rev Genet 7(10):793–803
84. Reik W (2007) Stability and flexibility of epigenetic gene regulation in mammalian development. Nature 447:425–432
85. Huang J, Fan T, Yan Q, Zhu H, Fox S, Issaq HJ, Best L, Gangi L, Munroe D, Mueqqe K (2004) Lsh, an epigenetic guardian of repetitive elements. Nucleic Acids Res 32:5019–5028
86. Li B, Carey M, Workman JL (2007) The role of chromatin during transcription. Cell 128:707–719
87. Guo H, Ingolia NT, Weisssman JS, Bartel DP (2010) Mammalian microRNAs predominantly act to decrease target mRNA levels. Nature 466:835–840
88. Katsnelson A (2010) Epigenome effort makes its mark. Nature 467:646
89. Barros SP, Offenbacher S (2009) Epigenetics: connecting environment and genotype to phenotype and disease. J Dent Res 88:400–440
90. Reik W, Walter J (1998) Imprinting mechanisms in mammals. Curr Opin Genet Dev 8:154–164
91. Okano M, Bell DW, Haber DA, Li E (1999) DNA methyltransferases Dnmt3a and Dnmt3b are essential for denovo methylation and mammalian development. Cell 99:547–557
92. Miranda TB, Jones PA (2007) DNA methylation: the nuts and bolts of repression. J Cell Physiol 213:384–390
93. Ansel KM, Lee DU, Rao A (2003) An epigenetic view of helper T cell differentiation. Nat Immunol 4:616–623
94. Zaina S, Lindholm MW, Lund G (2005) Nutrition and aberrant DNA methylation patterns in atherosclerosis: more than just hyperhomocysteinemia. J Nutr 135:5–8
95. Barker D, Eriksson JG, Forsen T, Osmond C (2002) Fetal origins of adult disease: strength of effects and biological basis. Int J Epidemiol 31:1235–1239
96. Razin A, Shemer R (1995) DNA methylation inearly development. Hum Mol Genet 4:1751–1755
97. Ito K (2007) Impact of post-translational modifications of proteins on the inflammatory process. Biochem Soc Trans 35:281–283
98. Herman JG, Baylin SB (2003) Gene silencing in cancer in association with promoter hypermethylation. N Engl J Med 349:2042–2054
99. Breivk J, Gaudernack G (1999) Genomic instability, DNA methylation and natural selection in colorectal carcinogenesis. Semin Cancer Biol 9:245–254
100. Lister R, Pelizzola M, Dowen RH, Hawkins RD, Hon G, Tonti-Filippini J (2009) Human DNA methylomes at base resolution show widespread epigenomic differences. Nature 462:315–322
101. Kerata MS, Botello ZM, Ennis JJ, Chou C, Chedin F (2006) Reconstruction and mechanism of the stimulation of de novo methylation by human DNMT3L. J Biol Chem 281:25893–25902
102. Feinberg AP, Tycko B (2004) The history of cancer epigenetics. Nat Rev Cancer 4:143–153